FUELING AMERICA

An Insider's Journey

JACK KERFOOT

ACKNOWLEDGEMENT

A special acknowledgement to my lovely wife who traveled the world with me and was always there to help me through the many challenges in my career.

I would also like to acknowledge my friends, who were kind enough to provide me input, feedback, and the encouragement to write this book, including Art Astarita, Kelvin Cates, Tina Cates, John Hughes, Randy Kubota, Mary Alice Murphy, and Robert Stafford.

TABLE OF CONTENTS

FIGURES

PREFACE

Oil is the lifeblood of the global economy. *Fueling America* describes an insider's journey in the oil industry, which began with the 1970s energy crisis. His journey took him around the world to small dusty towns, bustling cities, sweltering jungles and scorching deserts. The author describes a forty year career working with scientists, wildcatters, ministers, sheiks, tycoons and potentates in the quest to quench America's insatiable thirst for oil.

The author describes how thousands of small, dynamic oil and gas companies, not just the large companies met the challenge of the energy crisis. Creative and innovative minds were able to turn the 1970s energy crisis into an oil glut by the 1980s.

Fueling America also explains the evolution for each of the major types of energy. Coal and nuclear energy were once thought to be the answer to America's energy crisis. Today, the demand for coal and nuclear energy is rapidly waning. Wind turbines and solar panels were once thought to be expensive, impractical sources for power. Over the past decade, America has experienced dramatic growth in power generation from large wind farms and solar parks.

Countries around the world are now gravely concerned about greenhouse gas emissions and global warming. America has vast, undeveloped renewable energy resources which could dramatically reduce our country's greenhouse gas emissions and eliminate our dependence on foreign oil. The author firmly believes these renewable energy resources can and will be transformed into sustainable, reliable and cost effective energy to secure America's future.

CHAPTER 1

Kerosene Lamps Save the Whales

Since the dawn of time, mankind has faced an energy conundrum. Primitive man's conundrum was finding sustainable, reliable energy for life-giving fire. Fire gave our distant progenitors the opportunity to survive in a world inhabited with swift and powerful predators.

For thousands of years, biomass fuel (wood, peat, etc.) provided primitive man the only viable energy source. Ancient civilizations learned to harness wind for sailing ships and flowing water to power mills. However, biomass remained civilization's primary fuel for heat and light until the Industrial Revolution.

In the eighteenth century, forests were being rapidly depleted for fuel. Coal was discovered to have a higher heat value than wood and was also abundant. The invention and development of the railroad provided an efficient mechanism to transport large volumes of coal to major metropolitan centers. Coal was a sustainable and reliable energy source and answered mankind's initial energy conundrum. Coal replaced biomass as civilization's primary fuel source and slowed the devastation of the forests.

In 1859, George Bissell and Edwin Drake drilled the first successful oil well in the world in Titusville, Pennsylvania. The actual well was drilled to a depth of 69.5 feet, using a percussion or cable-tool drilling technique. Cable-tool drilling employed a lever to lift and then drop a heavy blunt chisel onto the rock, similar to the operation shown in *Figure 1*. The impact from the heavy, blunt chisel broke or fractured a few inches of rock. The fractured rock was then manually scooped out of the hole and the process was repeated.

Figure 1[1]

The location for the first successful oil well was selected simply on the occurrence of natural oil seeps in the streams and ponds in the area. The drilling objective was to remove the overlying rock and accelerate the natural seepage of oil to the surface. Once oil started to flow in the pit, the oil was put into tanks, transported, and sold.

Initially, the primary use of oil was the distillate, kerosene. Kerosene was a fuel that could be used in lanterns to light homes. Kerosene was significantly cheaper than other lamp fuels, like whale oil. In the 1860s, New England whaling had hunted the sperm whale to near extinction for its oil. The burgeoning oil industry effectively saved many species of whale from extinction.

The oil discovery in Titusville, Pennsylvania, set off a global search for oil throughout the world. Royal Dutch Shell drilled the first successful oil well in Indonesia in North Sumatra in 1885. Royal Dutch Shell then discovered oil in the jungles of eastern Borneo in 1898. American wildcatter Anthony Lucas discovered the giant oil field, Spindletop, in east Texas in 1901.

British Petroleum drilled the first successful oil well in Iran in 1908. Fossil fuels (coal, oil, and natural gas) provided the energy for the Industrial Revolution in Europe and the United States. In 1876, the development of the first gasoline-fueled vehicle created a new demand for oil. The economic growth from the Industrial Revolution further enhanced the demand for all fossils fuels in the world.

In 1911, the British Admiralty made a strategic decision to convert the fuel for

---

1   Photograph courtesy of the American Oil & Gas Historical Society (AOGHS.org)

the Royal Navy from coal to oil. Oil was more energy efficient and weighed less than coal. The lighter weight of oil meant the war ships of the Royal Navy could travel significantly faster than coal-powered vessels. Other navies of the world followed the Royal Navy, further increasing the demand for oil.

The rapidly growing demand for oil resulted in an explosion in the number of oil companies all over the world. Thousands of private sector oil companies were established from 1860 to 1950. This period of private sector growth resulted in a technology revolution in the oil industry. The percussion or cable-tool drilling techniques were replaced by sophisticated, rotary-table rigs, which enabled wells to be safely drilled thousands of feet below the subsurface. Enhanced oil recovery systems significantly increased the ultimate recovery of producing oil fields. Conrad and Marcel Schlumberger developed the first tools that could be run down the wellbore to measure rock properties and later the presence of hydrocarbons. In 1947, oil companies took their quest for oil into the ocean, drilling a well in the Gulf of Mexico. Gravity, magnetic, and reflection seismology surveys were developed to image the subsurface to identify possible oil fields. The quest for oil had gone from guesswork and gut feeling to the use of very sophisticated science and technology.

In 1941, the United States produced approximately 60 percent of the world's oil supply. By 1950, the United States produced 5.41 Million Barrels of Oil per Day (MMBOPD)[2] and consumed 6.46 MMBOPD a day.[3] The average price for oil in the United States in 1950 was $2.77 per barrel (42 gallons = 1 barrel).[4] The average price for gasoline in the United States in 1950 was $0.27 per gallon,[5] which is equivalent to $2.14 per gallon, when the price is adjusted to inflation (March 2015). In the 1950s, the United States became a net oil-importing country. However, global oil supply, demand, and competition kept oil and gasoline prices relatively flat.

In 1950, as the private sector growth period ended, the public sector or national oil company period began.[6] Many oil-importing countries formed public sector or national oil companies to secure the supply of hydrocarbons for their country's future growth. Many oil-exporting countries formed national oil companies to optimize their country's most valuable natural resource, oil. The global impact of the national oil companies would eventually surpass that of private sector oil companies like Exxon, British Petroleum, and Royal Dutch Shell.

After World War II, most scientists thought nuclear power was the energy of the

---

2   U.S, Energy Information Administration
3   Petroleum Products Supplied by Type 1949 – 2008, DOE, EIA Annual Energy Review
4   Historical Oil Price Chart, October 6, 2016, Tim McMahon
5   U.S. Office of Energy Efficiency & Renewable Energy, March 7, 2016.
6   "Evolution & Revolution of the E&P Industry," Jack Kerfoot, 2006 International AAPG, Perth, Australia

future. In 1945, the book *The Atomic Age* predicted homes, cars, and planes of the future would be powered by nuclear energy.[7] From 1960 to 1980, nuclear power experienced a global revolution in technology development and growth. The global nuclear energy capacity rose from 1 gigawatt in 1960 to 100 gigawatts in 1980.

In 1950, fossil fuels and nuclear energy appeared to be the solution to the world securing an unlimited supply of energy. Only twenty-five years later, the promise of nuclear energy would fade, and the world would once again be faced with another energy conundrum.

---

7   *The Atomic Age*, David Dietz, 1945

CHAPTER 2

Scrambled Eggs in Texas

My journey in the oil industry began in 1968. I started work as a correlator operator for a seismic contractor, Bendix GeoProspectors. A correlator operator is a technician on the seismic crew that is responsible for processing the recorded seismic data. The correlator operator will use specialized instruments to filter out noise and enhance the seismic signal. The processed seismic records were then sent to the oil company's geophysicist, who would use this data to select drilling locations.

Bendix GeoProspectors sent me first to Texas and then to West Virginia in 1969. They paid me the handsome wage of one dollar and five cents per hour, a whopping five cents above minimum wage. To their credit, they also paid me eight dollars per day for living expenses. In 1968, seismic crews worked ten to fourteen hours a day, seven days a week. For me, the long hours meant a great opportunity to save money for university, which was my primary goal.

Our seismic crew was contracted by an oil company to acquire seismic data. The oil company used the seismic data in their search for new oil fields. The main components in our seismic crew's tool kit included geophones ("jugs") and cables, the recording truck ("dog house"), vibroseis trucks ("shakers") and seismic data correlating equipment. The field crew planted the geophones ("jugs") in the ground and then connected each string of geophones to a cable. The cables were then connected to the recording truck ("dog house"). A technician in the recording truck, the "observer," confirmed that all the geophones had been connected and then signaled the vibroseis trucks (or "shakers") to commence the vibrating operations. The shakers were equipped with large metal pads, which were lowered and then vibrated against the ground.

Each vibration period lasted approximately twelve to fifteen seconds and was recorded on computer equipment in the dog house. The observer in the dog house then signaled the vibrators to stop. The observer then confirmed whether the data had been successfully recorded. The seismic crew then moved up to the next station and repeated the operation.

At the end of the day, the acquired seismic data was taken to the correlator operator, who processed the data to enhance the seismic signal and remove noise, such as vibrations from automobiles on nearby roads. The processed seismic data was then sent to the oil company's geophysicist to help him in his quest to find new oil fields.

The components in today's seismic crew's tool kit are virtually the same as 1968, although the equipment is far more sophisticated. Instead of correlators, modern seismic data is processed by powerful computers to enhance the data quality. The computers that process the seismic data today are far more sophisticated and powerful than the computers that launched America's first landing on the moon.

My job with Bendix GeoProspectors first took me to Mexia, Texas. The entire seismic crew would meet with the manager at 5:30 a.m., discuss the planned program for the day, and then leave for the field at 6 a.m. I would start with the field crew, laying out cables and geophones (hustling jugs). We would typically work until 6 p.m. I would then correlate the seismic records acquired during the day for the next three to five hours.

Although the hours were long, there were also good times. One of my favorite memories involves a bet I made at Spunky's Cafe, where the seismic field crew met for lunch. It began with a discussion of the recently released movie, *Cool Hand Luke*. In the movie, Paul Newman plays a southern convict on a chain gang who eats fifty hardboiled eggs on a bet. I commented, "Eating fifty eggs is impossible. I could only eat thirty-five eggs, and they would have to be scrambled."

My claim immediately elicited cries of "No you can't!" I was bombarded with numerous bets and long odds against achieving such a culinary task. Spunky's owner offered to charge me only ten cents per egg if I took the bet. He also offered that if I ate all thirty-five eggs he would give me a free breakfast every day for the next thirty days.

We all agreed that I had to eat all thirty-five eggs within forty-five minutes. My colleagues gave me great odds, ranging from ten to one up to twenty to one for every dollar I bet on this event. With visions of paying for my entire college tuition, I took the bet. We also agreed to hold the egg eating challenge that night at 9 p.m. after work.

News travels fast in Mexia, Texas. Throughout the day, I was accosted by the justice of the peace, deputy sheriffs, shopkeepers, and many more, who all wanted to get in on the action and bet against me. I gladly took on all comers. When I showed up at Spunky's Café that night, more than fifty people were waiting for the main event, the egg-eating competition.

The eggs came to me, twelve to a plate. I wolfed down the first two plates of eggs within fifteen minutes. The third dozen went down more slowly, but I finished off the last plate with five minutes to spare. My first year at university was paid in full thanks to *Cool Hand Luke* and my risk analysis of my egg-eating skills. Little did I know that success in the petroleum industry would also require skills in risk analysis.

In 1969, the seismic crew moved from Texas to West Virginia, where my company had a new contract to acquire seismic data for a natural gas company. This seismic program would give me another lasting memory, as well as an entirely new perspective of Appalachia.

Approximately one month into the program, our regular seismic observer was sick, and I filled in for him. We had just completed acquiring a seismic line and were preparing to move to the next seismic line in our program. When I opened the door of the recording truck, I found myself confronted by a man in coveralls pointing a shotgun at me! "Mr. Coveralls" immediately asked me, none too genteelly, if I was with the West Virginia Water Resource Department. Keeping a wary eye on the shotgun, I said no, I was not, and quickly explained why I was on his road and what I was doing. At first, I thought he had a whiskey still nearby and suspected I was with the Bureau of Alcohol, Tobacco, Firearms, and Explosives (BATF) agency.

Once "Mr. Coveralls" calmed down and lowered his shotgun, he explained that the West Virginia Water Resource Department had come to his home and had put red dye in his septic tank. He was angry because this dye had turned his well water pink! He didn't realize that this meant his septic tank was leaking into his well water. In those days, rural homes in many parts of West Virginia either had a septic tank or an outbuilding over the rivers. In 1969, it was hard to find a healthy fish in any stream in Appalachia. The state has made tremendous progress since 1969. However, I never drank well water from West Virginia again!

While in West Virginia, I received my draft notice. I left the seismic contractor and went into the U.S. Army and served one tour in Vietnam with the 101st Airborne Division.

In 1969, oil sold in the United States for an average price of $3.32 per barrel, equivalent to $21.81 per barrel, inflation adjusted (August 2016). The average price for gasoline in the United States in 1969 was $0.35 per gallon, or $1.77 per gallon inflation adjusted (March 2015). Even with the escalation of the Vietnam War, an abundant supply of oil kept gasoline prices relatively stable in the United States.

In 1969, the United States was producing 9.24 Million Barrels of Oil per Day (MMBOPD), while consuming 14.14 MMBOPD. The rate of foreign oil imports into the United States was rapidly increasing. The United States was developing an addiction to oil that could no longer be sustained by domestic oil production. In only four more years, the United States would experience its first "oil crisis." The energy conundrum for the United States had begun.

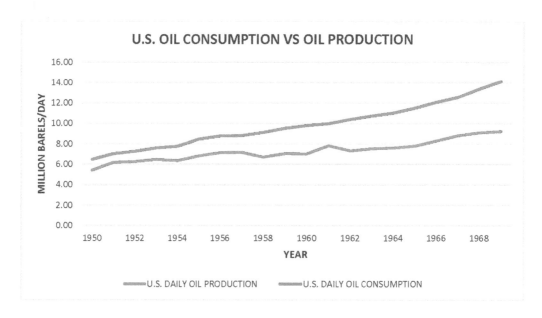

Figure 2

CHAPTER 3

The Sooners, Theory, and Practice

I received an Honorable Discharge from the U.S. Army and returned to the United States in November 1970, anxious to start university. Aided by the GI Bill, my savings from the seismic contractor, and my winnings from the egg-eating competition in Mexia, Texas, I enrolled at the University of Oklahoma ("Sooners") in January 1971.

As a baby boomer, I was profoundly influenced by the 35th President of the United States, John F. Kennedy. I believe, President Kennedy's words *"my fellow Americans, ask not what your country can do for you -- ask what you can do for your country"* still ring true, today.

In 1971, newspapers routinely described America's growing dependence on foreign oil. I felt that a career in the oil and gas industry would allow me another opportunity to serve my country. I only hoped that this time, America would be on the winning side.

As I began university, I was uncertain whether I should study to be a geologist, geophysicist or petroleum engineer? I thought if I gained as much practical experience in the oil and gas industry as possible, I would make a better-informed decision. In June 1972, I took a summer job with Sun Oil as a "roustabout" at their refinery, near Purcell, Oklahoma, approximately twenty miles south of the Oklahoma University campus.

As a roustabout, I did simple maintenance work, including painting the miles of pipes in the refinery. Although it was not overly challenging, the job did give me the opportunity to understand the fundamentals of processing oil into different products (gasoline, diesel, etc.). I also gained insight into the impact the Vietnam War had on me.

Approximately two weeks after I started work in the refinery, I heard a very familiar sound. I got up from the painting a pipe on the ground and pointed south and said: "In a few minutes, a formation of helicopters will be flying over the refinery. I expect there will be two Cobra helicopter gunships and five Huey helicopter troop transports." My coworker looked up and said, "You're nuts; I can barely hear you talk in this noisy refinery."

Two minutes later, seven U.S. Army helicopters flew over the refinery, including two Cobra gunships and five Huey troop carriers. In Vietnam, the sound of helicopters meant critical supplies, air support, or medical evacuation. I had been back from the Vietnam for more than a year, but that day I learned those memories would be with me forever.

Vietnam continued to play a role in my life at the most inopportune times. In July 1972, as I was driving my dark green 1968 Fiat 124 Spider sports car toward the not-so-bustling town of Purcell, Oklahoma, I passed an old pickup truck doing approximately twenty miles per hour in a thirty-five-mile-per-hour zone just as the sheriff of Purcell was approaching in the opposite direction. Sure enough, the sheriff made a U-turn and caught up to me. Just as I entered the Purcell city limits, the sheriff turned on his lights and pulled me over.

As was the practice in those days, I got out of my car with driver's license and registration, walked to the police car, and got in the passenger side of the police car, very, very slowly. I was greeted by the sheriff, who had a sawed-off shotgun pointed right at me. My first thought was, I got through Vietnam without a Purple Heart, and I really wanted to keep that record intact. My second thought was, this guy is treating me this way because he thinks I am part of the anti-war crowd at OU. After all, who but a hippie would drive a Fiat sports car around Purcell, Oklahoma? Keep in mind that in 1972 many Oklahoma residents believed the University of Oklahoma was a hotbed of radicals and hippies. I can attest to the fact that the rural residents of the state really didn't like radicals or hippies, full stop.

Fortunately, I was coming from work, so I was wearing coveralls and a hard-hat—not exactly hippie attire. And, as soon as I saw the shotgun, I switched into my "Okie" twang to let the sheriff know I was a true-blue Sooner. I was also careful to explain that this good ol' boy was also a veteran, who was headed home after a hard day's work at the refinery, up the road a piece. After mulling over the conflicting evidence, the sheriff grudgingly relented, "All right, boy, I will let you off, but make sure you don't get into any trouble in my town." I know this sounds like something from a 1970s Burt Reynolds movie, but that's exactly what happened. I never did return to Purcell, Oklahoma, but I can still switch on an "Okie" twang in a crisis.

In 1973, I got a summer job with Union Oil of California. I worked as a roustabout in the largest geothermal power plant in the United States in The Geysers, California, about seventy miles north of San Francisco. I saw this job as a unique opportunity to gain insight into renewable energy. At the time, the Union Oil geothermal plant provided over 80% of San Francisco's electrical power.

Although the roustabout's duties were about the same as in my previous summer job, the operating environment was dramatically different. Specially designed rigs drilled through hard granite rock, looking for fractures, which contained high-pressure steam. The high-pressure steam from the well was piped into a larger pipeline system that powered the turbines at the central electrical power plant.

Geothermal energy was thought to be the perfect energy to fuel electrical power plants. The process to produce geothermal energy created no pollution; the only by-product from the process was water from condensing steam. The water was then reinjected into the ground, which would be reheated by the earth to then become steam again.

Although California had rigorous environmental testing procedures, Union Oil did additional testing to maintain the highest possible environmental standards. That testing included analysis of the water composition from initial steam flow to its reinjection into the subsurface. During testing, it became apparent that the mineral content in the groundwater was increasing. The company notified the state, and then did additional testing. The tests revealed that the amount of water reinjected into the subsurface had to be greater than the amount of water from the condensing steam. Union Oil made the necessary changes and the mineral content in the groundwater returned to the normal level.

To me, this was an excellent lesson of the uncertainties in science and engineering. Continuous monitoring is always necessary to confirm that your theoretical assumptions are correct. In other words, Mother Nature is far more complicated than mere mortals can comprehend.

Outside of my little world, 1973 was a watershed year in the United States. In 1941, the United States was the largest exporter of oil in the world. By 1973, the United States had become the largest importer of oil in the world.

In 1960, the Organization of Petroleum Exporting Countries (OPEC) was formed by Saudi Arabia, Iran, Iraq, Kuwait, and Venezuela. OPEC's objective was to secure fair and stable prices for the petroleum-producing countries and a regular supply of petroleum to consuming countries. By 1973, OPEC consisted of eleven countries (Iran, Iraq, Kuwait, Libya, Qatar, Saudi Arabia, United Arab Emirates, Venezuela, Indonesia, Nigeria, and Ecuador).

In October 1973, Egypt and Syria invaded Israel to start the Yom Kippur War. The United States responded by providing emergency aid to Israel. The Organization of Arab Petroleum Exporting Countries (OAPEC), a subset of OPEC that supported Egypt and Syria, implemented an oil embargo against all countries that aided Israel. The United States responded to the oil embargo with price controls and rationing, which resulted in gasoline shortages throughout the country.

In 1973, the United States was producing 9.21 Million Barrels of Oil per Day (MMBOPD), while consuming 17.31 MMBOPD. Oil supply and pricing were now being driven by oil-exporting countries, many of which were hostile to the United States. The oil embargo forced the United States government to implement new energy policies. These policies focused on reducing dependence of foreign oil imports, providing incentives for domestic fossil fuel production, supporting development of nuclear power plants, establishing a "strategic oil reserve," and investing in renewable energy research and development.

The OAPEC oil embargo created acute gasoline shortages, frustration and fear across the United States. The oil embargo made me appreciate the importance of America's petroleum industry. I realized that a career in this industry would provide me the opportunity to help my country become less dependent on oil imports from the Middle East.

In June 1974, I accepted a summer job with Standard Oil of Indiana (AMOCO), working in the geophysical department in Houston, Texas. This was my first opportunity to work with experienced geophysicists who were exploring for oil and gas. The technical work, though elementary, was interesting. More importantly, I realized I had a passion for oil and gas exploration.

When I was growing up, Tulsa, Oklahoma, was considered the "Oil Capital of the United States." By the time I went to work for AMOCO, that title had passed to Houston. Houston was growing at unprecedented rates because of the rapid rise in the oil price and the prolific number of oil and gas fields in the Gulf of Mexico.

In June 1975, I was back in Houston, having accepted a summer job with Royal Dutch Shell in its international exploration department. My project used geological and geophysical data to predict subsurface pressures for offshore Sabah, Malaysia. I found this work exciting, since the results could be used to assess the oil and gas potential of the region. My supervisor and associates were experienced and highly proficient geologists and geophysicists. I felt drawn to international exploration in exotic places in the world.

On my return to university, I accepted a part-time position with Professional

Geophysics, Inc. (PGI), a specialty, high-end seismic processing company whose services were in high demand. PGI ran its seismic processing center twenty-four hours a day, seven days a week to try and meet clients' demands. This job provided excellent training in seismic processing, giving me skills few students in the United States would have upon graduation. Additionally, many seismic processing projects took several hours to complete, which left me with more time for my university studies. I truly enjoyed my university studies, whether it was geophysics, engineering, mathematics, or history.

In the spring of 1976, I got my first real opportunity to generate ready-to-drill prospects prior to graduation. I agreed to provide geological consulting services for an investor who lived in Texas. I was convinced that success at this job would further increase my marketability to potential employers.

The investor had a limited budget, which meant that and well would have to be relatively shallow. I started combing through the Oklahoma Geological Survey's files and found drilling records for wells drilled from 1910 through 1925 in northeastern Oklahoma. In those days, geologists kept a record of the rock types drilled and a qualitative description of any oil or gas that was encountered.

In the 1920s, natural gas had no economic value. As a result, all wells that encountered natural gas were immediately abandoned. I found several wells drilled from 1910 through 1925 that had encountered significant quantities of natural gas. Using the vintage drilling records and regional geological maps, I put together a portfolio of shallow, very low-risk natural gas prospects within a few months.

In 1975, natural gas was used in almost every home in the United States and natural gas pipelines crisscrossed the country. State utility companies were anxious to secure new natural gas reserves. In 1975, the price for natural gas was comparable to the price of oil.

I completed my report with drilling locations for five different prospects. I submitted my report and a detailed time sheet to support my rate of three dollars per hour to the Texas investor. After more than a month, I called to find out if there was a problem with my work. Many more calls followed over the next three months. Finally, I was paid. I suspect the investor was waiting to see if any of the prospects would make him money. I heard back from a friend that of the five prospects, three were economic successes, and the Texas investor received a substantial return on his investment.

Despite problems with payment, I gained tremendous job satisfaction and experience that increased my marketability to potential employers. I also learned a valuable lesson: not everyone will pay you in a timely manner, and only turn over the final product when you have been paid in full.

I was scheduled to graduate in August 1976 with a degree in geophysics. In the fall of 1975, I started writing letters to prospective employers and was pleasantly surprised by the quick and very positive responses I received from every company to which I had written. Perhaps I shouldn't have been surprised. The Society of Exploration Geophysicists (SEG) estimated that fewer than three hundred fifty people would graduate with a BSc, MSc or PhD in geophysics in 1976 in the U.S. and Canada, and less than 30% of these graduates would go to work in the petroleum industry. In 1976, the demand for geophysics graduates from the petroleum industry created a significant supply shortfall. I was very fortunate to have the right degree at exactly the right time.

In my career, I have had many interesting job interviews. However, one interview stands head and shoulders above the rest. In my final year at university, I received an academic scholarship from an oil company headquartered in Oklahoma. My scholarship advisor called and asked me to meet with his company's recruiter when he arrived on campus. Unfortunately, the company's employment recruiter had waited to visit my university one week prior to final exams. I had already received more than a dozen job offers from highly respected petroleum companies. Out of appreciation for the company that had given me the scholarship, I signed up for the interview with their campus recruiter.

I showed up at the appointed time and I met a gentleman who refused to make eye contact, gave me a dead fish handshake, and sat down behind the desk without saying a word. Campus interviews usually last only thirty minutes per person. After five minutes of absolute silence, I took the initiative and said, "Perhaps you would like to discuss my summer or part-time employment in the oil industry, or my thesis?"

The gentleman immediately looked up and said in an agitated voice, "This is my interview; I will let you know when I have a question." We sat for another fifteen minutes saying absolutely nothing. Finally, I said: "I noticed that you are focusing on one specific part of my resume. If there is anything you want to ask me, please feel free to do so." After a muted grunt, the gentleman said: "Do you date?" I thought he was joking, so I said: "Yes, would you like references?" His response, "How much?" At that point, I really had no idea what to say. I settled on, "Oh my, look at the time; I don't want to take up the next candidate's time." I stood up, shook his hand, thanked him for his time and the SEG academic scholarship, and quickly left the room. I was certain I would never hear from this company again.

The very next week, as I was beginning my final exams, I received a letter from the company saying, "Based on your campus interview, we would like to invite you to our office headquarters for an in-depth interview." My initial thought was, "Are you kidding me?" I politely declined in a neatly typed letter, which I mailed out the next morning. Within a week my SEG scholarship advisor called to tell me his company was very disappointed that I had declined their invitation. I said, "Based on the campus interview, I didn't think I would be a good fit for your company." I then gave a detailed account of the non-interview. My advisor said, "Our normal representative had to cancel. We brought this man in at the last minute. In his abbreviated orientation, we told him single people don't always enjoy working at our headquarters, as it is in a small town in Oklahoma." He then said, "I knew there had to be a good reason we were turned down for a headquarters visit by every student from Oklahoma University, University of Texas, Texas A&M, Tulsa University, and the Colorado School of Mines." I did learn a valuable lesson—you never know how someone will interpret your words unless you ask them!

I interviewed a total of eighteen companies on campus. These interviews, coupled with my work experience, helped me reach a better understanding of the diversity of company strategies and philosophies in the oil and gas industry.

Companies are sometimes referred to as "major" or "independent." While no exact definition exists, major companies usually have both upstream operations (exploration and production operations) and downstream operations (pipelines, refineries, and chemical plants), and tend to have organizational hierarchies, which can slow down their decision-making. Independent companies usually have more focused businesses, such as upstream operations in select geological plays or geographic regions, which usually leads to very efficient, nimble decision-making.

The major or independent distinction wasn't as important to me as the strategic differences between companies. Some companies focused their technical resources and capital investment on optimizing the oil recovery from existing producing fields. These companies placed a lower priority on finding new or future fields (exploration). Other companies focused their technical and capital investment on exploration. There were equally divergent opinions between companies on the development and application of new technology.

In the end, I received nineteen job offers from both major and independent companies. The demand was so high for geophysics graduates that I received an unsolicited job offer from one company with whom I never even interviewed. The criteria I developed to select a company included the quality of the technical training program for new employees, the company's geophysical expertise, the company's exploration focus, and the potential for future international assignments. Salary wasn't an issue, as all nineteen companies' salary offers were within six hundred dollars per year of one another. After reflecting on each company, I accepted a job with Mobil Oil, a major operator in Denver, Colorado.

In 1976, as I was preparing to start work in Denver, Colorado, the United States was producing 8.13 Million Barrels of Oil per Day (MMBOPD), while consuming 17.46 MMBOPD, as shown in *Figure 3*.[8] The average price for oil in the United States in 1976 was $13.10 per barrel. The average price for gasoline in the United States in 1976 was $0.59 per gallon, which is equivalent to $1.96 per gallon, when the price is adjusted to inflation (March 2015). In 1976, the United States rate of foreign oil imports was still increasing. The deficit of payments for foreign oil, coupled with the rapid escalation in the oil price, was having a devastating impact on the economy of the United States.

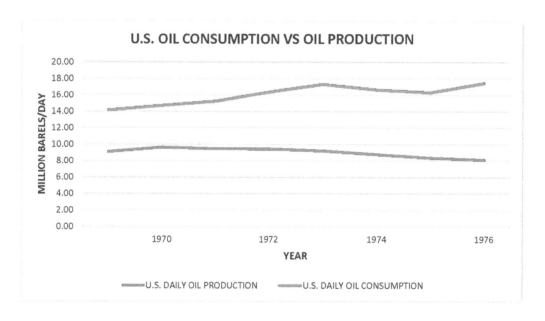

Figure 3

---

8   U.S. Energy Information Administration

CHAPTER 4

Rocky Mountain Highs and Lows

In August 1976, I drove into Denver, Colorado with all my belongings crammed into my now faded green 1968 Fiat 124 Spider and a grand total of sixty dollars to my name. I am sure my delipidated car, with my Oklahoma license plate, made me appear like a modern-day version of a character from *The Grapes of Wrath* by John Steinbeck. Fortunately, my new company had arranged for me to stay in a hotel on expenses for my first month of employment.

As I was preparing to turn into my hotel, a Denver police officer turned on his siren and lights. With flashbacks of the sheriff from Purcell, Oklahoma, going through my mind, I immediately pulled over to the side of the road. This police officer got out of the car then came up to me and asked to see my driver's license and registration. He then told me that one of my turn signal lights had burned out and he was going to write me a ticket. I immediately apologized for the taillight, and I told him why I was in Denver. I then told him I only had sixty dollars to my name, and that a new Fiat taillight cost over one hundred dollars. He said, "Look, I am writing you a warning ticket. You have thirty days to get the taillight fixed. If you don't have enough money to buy a new taillight in three weeks, I will write you another warning ticket, which will be good for another thirty days. I am almost always in this area from 5 p.m. to 8 p.m." I thanked him sincerely and thought, what a great welcome from the City of Denver!

Arriving at work the next day, I went through a brief orientation at Mobil Oil and was assigned a mentor. My mentor had been a tail gunner on a B-17 bomber flying during World War II. He had then acquired his PhD in geophysics with the GI Bill from the University of Nebraska. My mentor would oversee my technical training and development and, ultimately, become a good friend.

My mentor helped me learn the fundamentals of geophysical interpretation and hydrocarbon exploration. He allowed me to skip segments of the training program if I had previous work experience and I could demonstrate technical competency on the subject. As a result, I was able to move onto an exploration interpretation team within a few months. I could not have asked for a better mentor.

Within one month of starting work, I received a letter from the U.S. government on the repayment plan options for my student loans. I had taken out approximately four thousand dollars in student loans while I was at Oklahoma University. The letter from the U.S. government outlined the loan repayment options. Although there were over twenty repayment options, there were in fact only two options. The first option was for me to repay the loan within nine months, which meant I would pay no interest. The second option included multiple ways to repay the loan over an eight-year period with an 8% interest rate.

I am a frugal person. I didn't like the idea of paying interest on a student loan, and I also didn't like the idea of a debt hanging over my head for eight years. I decided to take a job as a bartender, working at night for the Celebrity Sports Center, which was owned by the Disney Corporation. The facility had indoor swimming pools, game rooms, bowling alleys, and restaurants. I worked twenty hours a week on the night shift. The real advantage of working two jobs is it is almost impossible to spend money, since all you do is work. Thanks to the Disney Corporation, I repaid my student loans in full within the nine-month period. I even received a letter from the Veterans Administration, stating I was the first Vietnam Veteran from the State of Oklahoma to repay his loan. My first thought, when I received this letter, was the Veterans Administration and the U.S. government had a significant student loan debt issue.

One of my first projects on the interpretation team was to determine if we could differentiate between a carbonate reef and mobile salt in the subsurface using seismic data. The study area was onshore Michigan. The company had several oil discoveries, but had drilled three straight noncommercial wells, called dry holes.

I spent many late nights and weekends running numerous geological and geophysical models to try and find a definitive answer. I found the project very interesting and complex. After four weeks, I was asked to give a presentation to the exploration manager and a team of senior technical professionals. Apparently, the exploration manager expected I would have completed two or three different models over a four-week period. I had in fact, completed over forty different geological models, using all the available geological and geophysical data.

After I finished the presentation, the exploration manager asked each person what they thought the probability was to use seismic modeling to differentiate between a prospective carbonate reef and a non-prospective mobile salt in the subsurface. Every one of the senior technical professions stated that there was an 85% to 95% probability of successfully predicting a carbonate reef by using the seismic modeling I had just presented.

The exploration manager then asked my opinion. I said there was only a 65% chance of successfully predicting a carbonate reef with the available seismic data. The exploration manager immediately asked, "Don't you have confidence in your work?" I responded, "Yes, I have a very high level of confidence in my work; however, I also know the uncertainty there is in the analysis." Mobil Oil drilled the well and found salt and abandoned the well as a dry hole.

In any evaluation, it is essential that you know the range of uncertainties, or know what you don't know! My assignment was my introduction of quantitative decision and risk analysis, which would become a fundamental form of analysis I would need throughout my career.

After this project, I was assigned to the exploration team working Alaska. The largest oil field in the United States, Prudhoe Bay Field, is located on the northern slope of Alaska. The exploration team worked up a prospect east of the Prudhoe Field and drilled a dry hole.

At this point in time, I felt like a professional failure. While I was in university, I had gotten five prospects drilled and three of the wells were economic successes. I had been working for Mobil Oil for two years and had done technical work to support two exploration prospects, which were both economic failures.

Several of my experienced colleagues told me they had never had an exploration well drilled in their ten-plus years at Mobil Oil! I also found out that the company only drilled three exploration wells a year in the entire region, which included the entire western U.S. and Alaska. I was flabbergasted, as I couldn't understand how any company stayed in business without finding new oil reserves.

I asked our exploration manager about the economics of our regional exploration program. He told me that I was the first person to ask this question. Surprisingly, he didn't have an answer for me, but just said, "Good question." At the same time, I was getting telephone calls from an independent oil company, Forest Oil about joining their organization.

I flew down to Midland, Texas, and met with the president and exploration manager of the company. They told me I would have an opportunity to drill three or more wells a year if I joined them in Midland, Texas. They further promised to provide me a series of advanced technical training courses to make sure I continued to develop my technical skills. Midland, Texas, is not Denver, Colorado, when it comes to quality of life. However, I saw this as an opportunity to really grow professionally. I accepted the job offer, flew back to Denver, Colorado, and submitted my resignation to Mobil Oil.

Within minutes of submitting my resignation, the exploration manager came to my office and asked why I had resigned. He asked if I was leaving for financial reasons and I said, "Absolutely not!" I told him that although I was getting a 20% pay raise, I was leaving for the opportunity to drill exploration wells and further my professional development.

The exploration manager then asked, "How can you resign after receiving an exceptional performance appraisal?" My response, "What is a performance appraisal?" Apparently, the supervisor of my exploration team had failed to tell me about the appraisal performance process and also failed to hold the appraisal discussions with me. My supervisor just submitted the form without a formal review, because he thought that I must know I was doing good work. Unfortunately, he didn't realize that I had failed Mind Reading 101.

In 1978, as I was preparing to leave Denver, Colorado, the United States was producing 8.71 Million Barrels of Oil per Day (MMBOPD), while consuming 18.85 MMBOPD a day. The average price for oil in the United States in 1978 was $14.95 per barrel. The average price for gasoline in the United States in 1978 was $0.63 per gallon, which is equivalent to $1.83 per gallon, when the price is adjusted to inflation (March 2015). In 1978, the United States rate of foreign oil imports was still increasing. The deficit of payments for foreign oil, coupled with the rapid escalation in the oil price, was continuing to have a very negative impact on the economy of the United States. The ever-increasing price of oil and gasoline did nothing to thwart the United States' oil addiction.

CHAPTER 5

Texas Tea and the Canadian Cold

In the summer of 1978, I drove from Denver, Colorado, to Midland, Texas, to start work for Forest Oil. It quickly became apparent that my new company had very different philosophies and strategies from my previous company. My former company was considered a major, while my new company was considered an independent. However, the differences between the two companies were much greater than corporate size.

Forest Oil was established in the early 1900s. The company developed a new technology to increase the amount of oil recovered from a field. The new technology, water flooding, allowed companies to increase the amount of oil recovered from a field by 10% to 20%. After World War II, the company focused on oil and gas exploration in the United States.

Mobil Oil focused primarily on the application of technology in finding and developing oil and gas fields. The work pace was measured and at times I felt like I was working in academia. Forest Oil was focused on value, which meant a balance of technical and commercial factors.

Another significant difference between the two companies was the compensation for the technical professionals. Mobil Oil's compensation system was based on the person's performance and salary grade. Forest Oil's compensation system was based on a person's performance, salary grade, and a royalty program. The company's royalty program allowed the technical professionals to buy up to a 4% royalty in any exploration prospect they generated. If the prospect was a dry hole (uneconomic), then the invested employees and the company both lost money. A 1% percent interest in one exploration well could cost the technical professional two thousand to five thousand dollars.

If the prospect was a success, then all the investors reaped the rewards. As an example, an exploration discovery with gas reserves of one hundred billion cubic feet (BCF) would generate a revenue of approximately one million dollars over the twenty-year life of the field. In 1978, the commercial success rate was only 10% to 15% for all drilled exploration prospects in the world. Forest Oil's royalty program ensured only the best prospects were drilled.

Midland provided me with a very different lifestyle than Denver. The oil business wasn't the primary industry in the "Mile High City" of Denver. The outdoors and sports were a high priority for many of the people who lived in Denver. In Midland, almost everyone was fixated on the oil and gas business. I saw people discussing oil or gas investment opportunities in a McDonald's restaurant! It was virtually impossible to talk to anyone in Midland without the topic of the oil and gas business entering the conversation.

The rapid increase in oil and gas prices from 1972 to 1978 created a boom in the industry. Unfortunately, it isn't uncommon to find criminal activity when there is a boom in any business. Shortly after I arrived at Forest Oil, all the employees were called into the conference room and told the Federal Bureau of Investigation had just made arrests of an oil field equipment theft ring. The criminals were stealing oil field equipment from the Midland, Texas area and reselling the equipment in other regions of the United States. This was quite an eye-opening experience for me.

My most memorable, non-work-related experience in Midland was the live performance of "Play It Again, Sam" at the local dinner theater. Bob Denver, who played Gilligan in the television comedy, *Gilligan's Island*, was the lead actor in this play. The theater stage was on an open circle, which was surrounded by three levels. Each level had dinner tables, which looked down on the theater stage. The dinner theater held approximately one hundred people.

When I arrived at the dinner theater, I found most of the theater guests were at the bar consuming large quantities of amber liquid (whiskey). Once the play started, it was obvious the consumption of alcohol had not slowed down one iota. Approximately halfway through the play, Bob Denver came out center stage to deliver a soliloquy in the play. He paused to see more than a third of the audience passed out, face down on their tables. Bob Denver simply shrugged his shoulders and proceeded to deliver his lines. I was very impressed with the actor's professionalism. I also learned that that night, I may be faced with a difficult situation, I can certainly control the way I react to the situation.

When I joined Forest Oil, the plan was for me to work in the regional office in Midland, Texas, for approximately six months and then transfer to the Oklahoma

City, Oklahoma, office. After six months, I found myself once again driving with all my possessions in my car to my next job assignment.

The Oklahoma City office had been open for less than two years. The office had no oil or gas production and therefore no cash flow. I quickly learned that cash flow is a critical factor in maintaining an office, as well as employment with a company.

The Oklahoma City office had been drilling very deep (twenty-five thousand feet to thirty thousand feet), high-risk, and high-potential exploration wells in the Anadarko Basin in central Oklahoma. A single well would cost from twenty-five to forty million dollars to drill. The Forest Oil Corporation could only afford to fund a 25% in one of these deep Anadarko Basin wells per year.

Forest Oil would form a consortium with four or five other companies to drill one of these expensive, high-risk, and high-potential wells. Each well took over three hundred days to drill and the commercial success rate for these exploration wells was only 10%.

My first project in the Oklahoma City office was to determine if our company should continue to invest in these expensive, high-risk, and high-potential exploration wells in the Anadarko Basin. This type of evaluation would require me to quantify the value of all the major exploration plays in the Anadarko Basin and Arkoma Basin in Oklahoma. It was obvious to me that this type of analysis should be done. However, I had never done anything remotely like this analysis in my limited professional career. Fortunately, I found many good books on exploration play analysis and assessment at the University of Oklahoma library.

My assessment estimated the number of prospects and the reserve potential of a typical prospect in each geological play. Historical drilling data provided me with the commercial success rate for each geological play. I ran economics on each play, which allowed me to quantitatively rank all the plays in the Anadarko and Arkoma basins.

After two months of intense work, I presented my analysis of over fifty geological plays to Forest Oil's senior management. Two geological plays offered significantly greater value to Forest Oil than the high-risk and high-potential geological plays the company had been drilling. I also presented several exploration prospects in one of the two high-rated plays. Forest Oil's senior management gave our office the approval to proceed with this new exploration program. It was an exciting time for a young exploration geophysicist like me.

Forest Oil acquired the mineral leases for one of the high-rated prospects in the new exploration play. The United States is one of the few countries in the world where private citizens can own oil and gas mineral rights. In 1979, a typical mineral rights agreement would include an annual leasing fee, a royalty on any oil and gas produced from the land, and payment for any surface or subsurface damages to the landowner. At the time, the annual leasing fee was in the one hundred to three hundred dollar per acre range. The typical royalty in an oil and gas agreement was 25% of the revenue for the gross production. In Oklahoma, oil and gas mineral interests have generated billions of dollars of revenue to the mineral rights owners.

A second prospect in the new play was held by a producing field, operated by Royal Dutch Shell. The field was producing from a shallower reservoir, thousands of feet above the exploration prospect that I had identified. Public records showed that Royal Dutch Shell's field was no longer producing natural gas. Royal Dutch Shell was required to either reestablish oil or gas production within sixty days or relinquish the lease back to the farmer who owned the mineral rights.

Forest Oil decided to "top lease" Royal Dutch Shell's existing oil and gas lease. If Royal Dutch Shell could reestablish production, Forest Oil would lose the money paid to the farmer for the "top lease," which was approximately one million dollars. If Royal Dutch Shell couldn't reestablish production within sixty days, then Forest Oil would have the mineral rights for the lease and we could drill the exploration prospect.

Most companies have a "scout," who is a person that checks out what other companies are doing in your areas of interest. Scouts have a wide range of duties and are the equivalent to a company spy in the oil and gas industry. Our scout determined that Royal Dutch Shell was drilling to a deeper target, which was almost certainly the same prospect our team had identified.

Everyone in our office hung on every report from our scout. Finally, we received word: time had run out on Royal Dutch Shell. Forest Oil now owned the mineral rights permit over our prospect in the new exploration play. However, Royal Dutch Shell was responsible for the abandonment of the drilled well and remediation of the land prior to exiting the permit.

Our office contacted Royal Dutch Shell and agreed to take over the rig and the well-abandonment program. This would save Royal Dutch Shell tens of thousands of dollars in abandonment costs for the well that had just been drilled. However, Forest Oil now had to drill only four hundred feet deeper to test the exploration prospect, which would save our company millions of dollars in drilling.

Forest Oil's drilling department took over the rig operations from Royal Dutch Shell, and a few weeks later we had a commercial gas discovery. This was the first commercial discovery for our Oklahoma City office. It was a day of celebration for everyone in our office and our company.

Unfortunately, Forest Oil had decided to change the royalty program for technical professionals. In the new royalty program, I could no longer invest my own money in a prospect that I developed and share in the financial reward if the exploration well was a discovery. Forest Oil had decided to start a "pooled override program," which would include all salaried employees in each of the company's seven offices. When this change was announced, I was told, "Don't worry, we expect this program will equal your annual salary in the next two to three years." At the time, I was paid less than $30,000 per year.

The discovery in our Oklahoma City office was put onstream within a few weeks and everyone in the office monitored the daily production reports. In 1979, natural gas in Oklahoma was selling for more than eight dollars per thousand cubic feet. This was one of the very few times in the United States that the British Thermal Unit (BTU) price for natural gas was higher than the BTU price of oil.

Four months after our exploration discovery began producing natural gas, I received a royalty check from Forest Oil. My hands were shaking as I opened the royalty check, as I had no idea what size bonus I might expect. After I opened the envelope, I saw a check for only five cents. To add insult to injury, at the bottom of the check was the statement: "This check is not valid for amounts less than one dollar." Five cents! I couldn't even buy a hamburger at McDonald's for five cents.

I immediately went to my manager, who was staring in disbelief at his royalty check. He looked up from his desk and said, "Yes, I know you are surprised. Don't worry; I know there must be a mistake." Unfortunately, there was no mistake with the royalty calculation. The issue was the royalty on our exploration discovery was being split with several hundred salaried employees in the seven Forest Oil offices. The new royalty program weighted each person's share by grade and time of service with the company. The accounting manager in the Jackson, Mississippi, office received a significantly larger royalty check than the entire team that generated the exploration discovery in the Oklahoma City office.

Unfortunately, Forest Oil turned a highly motivating incentive program into a disincentive. The company started to lose many of the very talented technical professionals that had helped the company grow and prosper over the previous decades.

In the fall of 1979, I was asked if I would consider a temporary assignment in the company's Calgary, Alberta, Canada, office. Although I had just started law school in the evening, I thought this might be an excellent opportunity to see new exploration geology and also live in a different country. This decision would forever change my life for the better.

The prime minister of Canada was Mr. Pierre Trudeau. The prime minister was a nationalist, who opposed global trade and foreign workers in Canada. The prime minister's policies made it very difficult for any foreigner to get a work permit in the country. These policies created a severe shortage of oil and gas technical professionals in Canada. During this period, it was not uncommon for experienced technical professionals to change companies three to four times a year. The technical professional would then receive a twenty to thirty percent salary increase from the new company.

Forest Oil was required by Canadian law to advertise my position for six months in all the major newspapers. My American salary was significantly below average for a Canadian with my work experience. As a result, no Canadian applied for the position and I was granted a three-month work permit. Forest Oil had to reapply for my work permit every month to extend my stay in Canada.

I landed at the Calgary airport in September 1979 and quickly learned the meaning of "cold weather" when I walked outside the airport to locate a taxi. I realized I would have to buy warm, Canadian clothes, or I would soon become a human Popsicle.

Driving in Calgary during the winter was also a unique experience for me. I had to learn how to drive through very icy streets. I also had to learn what the strange electrical plug was doing in the front of my company car. Kind Calgarians explained that the electric plug on my car was a heater to keep my engine block from freezing and cracking during the bitter cold winters.

Calgary is a beautiful city, situated at the base of the Rocky Mountains. The people in the office were very nice, but it was a very laid-back office. The office opened at 8 a.m., but my coworkers didn't start filing into the office until 9 a.m. The critical shortage of qualified technical professionals in Calgary certainly contributed to the laid-back work atmosphere.

One of my responsibilities was to set up and train the staff in the use of a new geophysical workstation system. I met with the building manager and gave him the wiring specifications for the new workstation. I was assured there would be no problem. In fact, it took many long hours and late nights to determine that the wiring specifications for the building were incorrect. When I confronted the building manager about this inexcusable oversight, his response was, "Oh well, things happen." I suspect he didn't appreciate my vociferous response.

The geology in western Canada was very interesting. However, it was obvious there was a lack of due diligence in assessing the geological risk of the exploration prospects. In Canada, the federal government (Crown) owns all mineral rights. The Crown holds regional lease sales, several times a year. Companies submit sealed bids and the highest bidder is awarded the oil and gas rights for a number of years.

The Calgary office submitted an aggressive bid of more than fifteen million U.S. dollars for the mineral rights on a license in Alberta, Canada. Unfortunately, the technical team that generated the exploration prospect on this very expensive lease overlooked critical technical data. The exploration well was a multimillion-dollar dry hole. A simple post-drill analysis caused numerous red faces, although no one seemed overly concerned except me.

While I was working in Calgary, I was contacted by several industry recruiters (headhunters) about job opportunities all over the world. Headhunters are paid significant commissions by companies to fill critical positions. I was also becoming concerned about Forest Oil's financial stability. I wasn't convinced that Forest Oil would survive if the oil and gas prices retreated from the current record highs. I began interviewing with companies for overseas job opportunities.

I didn't fully appreciate the economic forces that caused the rise in the oil and gas prices. I simply thought that whatever goes up will certainly go down at some point in time. In a few short years, I would see firsthand the devastating impact of plummeting prices on the industry.

While I was interviewing with different companies, I also fell in love with a beautiful Canadian woman. In a very short time, I had accepted a position with another company and had gotten married. However, I didn't resign from my position with Forest Oil until I was sure my wife, Barbara, could legally enter the United States. One of my more memorable life experiences was my dealings with the United States Consulate office in Calgary, Alberta, Canada.

Barbara and I went to the U.S. Consulate office to apply for her U.S. entry visa. After many hours, we finally met an embassy representative who told U.S. it would take up to a year for Barbara to get an entry visa. I didn't argue, but this information was inconsistent with what I had previously read on U.S. immigration policy.

Fortunately, a close friend of the family was an Oklahoma congressman. I called the congressman and he told us to return to the U.S. Consulate that next day at 10 a.m. Barbara and I arrived at the U.S. Consulate at the designated time and were immediately greeted by the consulate general. The consulate general went out of her way to help us through the reams of paperwork that had to be completed to get

Barbara an entry visa to the United States.

Barbara was instructed by the U.S. Consulate to take a comprehensive medical exam in addition to documenting her entire life story. After one week, Barbara and I returned to the U.S. Consulate and submitted the paperwork. Barbara received her visa within a few days, thanks in no small part to the congressman.

Once we had the visa, I submitted my resignation, which was met with surprise in the Calgary, Oklahoma City, and Midland offices. I stayed in Calgary several weeks to fulfill my notice period and to tie-up loose ends at work. I wanted to leave Forest Oil on good terms.

Barbara and I drove out of Calgary with all our possessions in our car. After many hours of driving on icy winter roads, we finally arrived at the Canadian and U.S. border. One lone U.S. immigration agent manned the small border station in Idaho. Barbara and I explained our purpose of business and then presented our passports with Barbara's voluminous visa documents.

The immigration agent stamped Barbara's passport and asked, "Do you want any of these documents?" When Barbara said, "No," we were shocked to see the immigration agent throw all of her entry visa documents into a large trash bin. The immigration agent explained that the U.S. Immigration Agency had been sending numerous letters to the U.S. Consulate Office in Calgary, explaining the paperwork was no longer required!

When we ultimately arrived in San Francisco, we found out Barbara could have simply entered the U.S. on a tourist visa and then applied for a green card, since she was married to a U.S. citizen. Please count me as one of many who believe the United States Immigration Policies should to be overhauled.

As Barbara and I were leaving Canada, the United States was producing 8.71 Million Barrels of Oil per Day (MMBOPD), while consuming 18.51 MMBOPD a day. The average price for oil in the United States in 1979 was $25.10 per barrel. The average price for gasoline in the United States in 1979 was $0.86 per gallon, which is equivalent to $2.31 per gallon, when the price is adjusted to inflation (March 2015).

Over the next two years, rising oil prices caused a decrease in oil consumption in the U.S. The oil and gas industry responded to high oil prices by significantly increasing exploration drilling. The increase in exploration drilling resulted in an increase in oil and gas production in the U.S. This phenomenon was seen around the world.

In just a few years, the decrease in oil demand and the increase in oil production resulted in a dramatic drop in oil prices. However, economic "experts" were projecting oil prices to climb to over one hundred dollars per barrel within the next three years. The economic "experts" simply ignored the supply-and-demand side of oil in their pricing forecasts.

While the oil and gas industry was booming, growth in the nuclear energy industry was slowing. In the 1970s and 1980s, the cost to build nuclear power plants increased significantly due to extended construction times that were associated with new government regulations from nuclear power plant incidents and growing opposition from anti-nuclear energy organizations.

In the late 1960s, some scientists began to express concerns about nuclear energy, including nuclear waste disposal and potential nuclear plant accidents. These concerns developed into a global anti-nuclear movement. The debate on nuclear energy was intense and resulted in a measured slowdown in the construction of nuclear power plants in Europe and North America.

CHAPTER 6

Jungles of Borneo

After evaluating several job opportunities, I accepted a position with HUFFCO, an independent oil and gas company with a major operation in the Republic of Indonesia. The opportunity would provide me with international exploration experience, which was one of my professional goals. Barbara and I also met several people who spoke highly of their time living in Indonesia and of the wonderful Indonesian people. We thought Indonesia would also give us the opportunity to travel and see some of the wonders of Asia.

Barbara and I did our research on the Republic of Indonesia prior to accepting the position with HUFFCO. Indonesia had been a Dutch colony that gained independence in 1945. In 1980, the president of Indonesia was the political strongman, Suharto. President Suharto had seized power, following a failed coup d'état attempt by the Indonesian Communist Party in 1965.

The government under President Suharto was infamous for corruption and favoritism. In 1980, the Republic of Singapore and the Republic of Indonesia were in a dispute over the location of the probate of an Indonesian citizen who had passed away while living in Singapore. This man had been the number-two man in the Indonesian State oil company, PERTAMINA.

The dispute was in all the newspapers, because the former PERTIMINA manager had amassed over twenty-five million U.S. dollars in accounts in various Singaporean banks. The man's widow was asked on television how her late husband had amassed such a fortune while making only five thousand U.S. dollars a year from PERTAMINA. His window's response: "He was a very thrifty man." I met several "thrifty men" in my time in Indonesia.

Indonesia is located along the equator between the Indian Ocean and the Pacific Ocean, as shown in *Figure 4*. The country is known as the land of a thousand islands. In 1980, the population of Indonesia was approximately 145 million people. Approximately sixty percent of the people lived on the island of Java. Many of the regions, like Kalimantan, Sulawesi, and Maluku were sparsely populated.

Indonesia has approximately three hundred different ethnic groups. However, the largest ethnic group in Indonesia is the Javanese, who compromise approximately forty percent of the country's population. The Javanese dominated all the key positions in the Indonesian government. Most of the people that live on the islands of Java and Sumatra are Muslim. Indonesia has the largest population of Muslims in the world.

Figure 4

Prior to joining HUFFCO, Barbara and I received an orientation at the company's headquarters in Houston, Texas. The company's Indonesian office was located near the town of Balikpapan in East Kalimantan on the Island of Borneo.

In the orientation, we were told we would live in a company compound. In the presentation, the company compound looked like a country club, complete with tennis courts, swimming pools, racquetball courts, bowling alleys, and a commissary. Barbara and I would live in an American-style apartment, complete with all the modern conveniences, including a refrigerator, washing machine, and air conditioning.

In January 1980, we flew nonstop from San Francisco, California, to the Republic of Singapore. We spent the night in Singapore, and then returned to the Singapore International Airport the next morning to fly to our future home in Balikpapan, Indonesia.

There was only one flight per day from Singapore to Balikpapan on a small commercial jet. At the gate we met a woman whose husband was the finance manager for HUFFCO. It turned out she and I had both grown up in the same town of Tulsa, Oklahoma.

As we got on the plane, Barbara quietly asked me, "Why don't you have an accent like hers, since you are from the same town?" I explained to my bride that when I was sixteen years old I attended a National Science Foundation summer program in earth sciences in Corvallis, Oregon. I was the only person from the southern part of the United States. At that time, I had a pronounced "Okie" accent. Because of my accent, I was mercilessly teased by the other students in my class. Upon my return, I made a concerted effort to lose my accent. I would watch Walter Cronkite on the evening news every night to learn how to enunciate my words and rid myself of my "Okie" accent. This explanation was met with a pronounced "good" from my bride.

Our flight from Singapore to Balikpapan International Airport took approximately two hours. Balikpapan International Airport consisted of a tarmac landing strip, no control tower, and a one-room shack where passports were stamped, and visas checked. It took less than fifteen minutes to disembark from the airplane, collect our luggage, and clear the Indonesian customs.

There were no hotels in Balikpapan, Indonesia. Barbara and I were driven directly to the company camp and immediately moved into our new apartment, as advertised in our company orientation. My Canadian wife immediately set the air conditioning thermostat to subarctic temperatures, where it stayed for our entire time in Balikpapan! Anyone entering our home would have to wear a sweater or risk frostbite!

The next morning, I caught the company bus to the office. The distance from our camp to the company's office was approximately six miles. However, the trip took thirty to forty minutes due to the size and frequency of potholes on the road. Our daily trip to the office passed by the bustling local market. On hot, windless days you could smell the pungent aromas from the market several blocks before we passed it. I have many vivid memories of the bouncing morning bus rides to our office.

HUFFCO's oil and gas license in Indonesia was a large region in swamps and jungles of East Kalimantan on the island of Borneo. The west-to-east flowing Mahakam River, as shown in *Figure 5*, bisected HUFFCO's oil and gas license. The company had already discovered two giant gas fields (Badak Field and Nilam Field) in the delta of the Mahakam River. The giant gas discoveries had resulted in the building of a large liquefied natural gas (LNG) plant, approximately one hundred forty-five miles north of Balikpapan, near the town of Bontang.

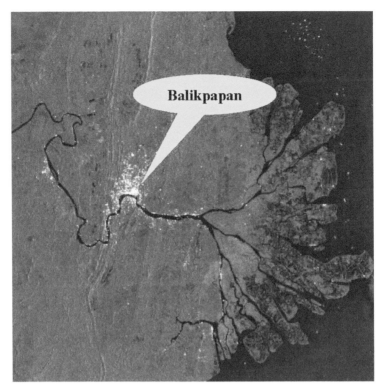

Figure 5[9]

Most of HUFFCO's drilling and seismic operations were either in dense jungle, as shown in *Figure 6,* or in the swamps along the Mahakam River.

Figure 6[10]

---

9   Satellite photograph by Earth Snapshot
10   Courtesy of Mr. Art Astarita

East Kalimantan has an equatorial climate with average daily temperatures ranging from 80 to 90 degrees Fahrenheit.[11] Balikpapan doesn't have a classic wet and dry season. The average monthly rainfall in East Kalimantan ranges between six to nine inches. As a result, the daily humidity ranges from 80% to 90% throughout the year. The climate and the Nipa Swamp made drilling and seismic operations very difficult.

Seismic operations in East Kalimantan, Indonesia, were dramatically different from the onshore operations I had experienced in the United States and Canada. The onshore seismic crews in the United States and Canada had between thirty to forty people on each crew. The seismic operations in the jungles of East Kalimantan had between 2,000 to 2,200 people on each crew.

Each jungle-portable seismic crew had a survey team, a shot-hole team, and a recording team. The survey team measured and cut the trail for each seismic line. The survey team would also construct landing pads for the helicopters, which would resupply each jungle-portable seismic crew, as shown in *Figure 7*.

Most of the onshore seismic crews in the United States and Canada use vibrators, mounted on large trucks, to shake or vibrate the ground. The vibration creates a shaking motion like a mini-earthquake. The vibration creates a signal, which the geophones on the ground pick up and then transmit to the recording truck. The seismic signal is then recorded on a computer in the recording truck.

In HUFFCO's jungle-portable seismic operation, the energy source was plastic explosives. The shot-hole team would drill and case a hole between sixty to one hundred feet in the ground, as shown in *Figure 8*. The plastic explosive would then be positioned with an anchor at the base of a shot hole.

---

11   World Weather & Climate Information

Figure 7[12]

Figure 8[13]

---

12   Courtesy of Aled Morgan
13   Courtesy of Aled Morgan

After the shot-hole team set and secured the plastic explosive charge down the shot hole, the recording team detonated the charge and recorded the seismic data. The seismic data was recorded on a sophisticated jungle-portable computer system.

The jungle-portable seismic crews in HUFFCO's operation acquired sixty to ninety miles of seismic data per month. The seismic data was shipped to the seismic processing center in the Republic of Singapore. It took the seismic processing center approximately six to eight weeks to process each seismic line. The final processed seismic lines were then shipped to me, the geophysical interpreter in Balikpapan.

The final processed seismic data allowed the geophysical interpreter and the geologist to develop a detailed picture of the subsurface. The team's objective was to locate "good" drillable prospects. The term "good" simply means the prospect had the potential to provide a significant return on the investment, including the costs for seismic data, drilling wells, overhead, etc.

In 1980, development prospects typically had an 85% to 90% chance of commercial success. Exploration wells typically had a 10% to 15% chance of commercial success. However, the exploration prospect size was orders of magnitude greater than a development prospect. As with any business, the bottom line was about managing risk and reward to provide the shareholder with a positive return on his investment. Failure to understand this basic principle would quickly drive a company into financial ruin.

In 1980, HUFFCO was pursuing a very aggressive exploration and development program. The company was operating eleven drilling rigs and three jungle seismic crews with less than a dozen technical professionals (drilling engineers, reservoir engineers, geophysicists, geologists, etc.). When I walked into my office, I found over three thousand miles of seismic data that urgently needed to be interpreted. My boss's only guidance was, "Find good drillable prospects, quickly."

HUFFCO had a small team of exceptional expatriate technical professionals working on the development of the two giant fields and the exploration program. Gas from the ongoing development of the two giant fields was transported from the wellhead by pipeline to the LNG plant in Bontang. The gas from the LNG plant was then sold to buyers on the global market, primarily Korea and Japan. Once the LNG was sold, the liquefied natural gas was transported to the buyer in giant, refrigerated tank ships. Exploration's goal was simple—find new commercial quantities of oil or gas, as quickly as possible.

An oil company must pay the drilling company a daily rate to use the rig, including the time to mobilize the rig into the region and to demobilize the rig out of the region. Oil companies try to contract a drilling rig for an extended period of time to minimize the mobilization and demobilization costs. The oil company's challenge is to keep the rig continuously drilling over the length of the drilling contract. An inventory of quality prospects is essential to keep a drilling rig operating efficiently during the length of the drilling contract. If you run out of prospects, the company must still pay for the drilling rig, even if it is standing idle. "Feeding" the hungry drilling rigs is always a challenge for oil and gas companies throughout the world.

At HUFFCO, we had eleven very hungry rigs devouring our prospect inventory at an alarming rate. My boss's guidance was on the mark. If I couldn't quickly find several good drillable prospects, one or more of our rigs would be idle. Fortunately, HUFFCO had three experienced exploration geologists who provided me a crash course in the geology of East Kalimantan.

After the exploration tutorial, I completed an in-depth study of Badak and Nilam, the two giant natural gas fields in HUFFCO's license. The field studies provided me valuable insight into the producing reservoirs, which helped in my hunt for quality prospects. After many late nights, I developed a new exploration play, which contained several exploration prospects.

The new exploration play was designed to drill discrete sand packages on a large, elongated structure (anticline). I was relatively confident that I could map the sand packages on the high-quality seismic data. In my geological model, each of the sand packages would be encased or sealed by shales. The sand packages would be charged with gas, which would migrate up a large bounding fault. The major risks in this exploration play were the sealing capacity of the shales and the ability for gas to migrate into each of the discrete sand bodies.

The anticline was approximately thirty miles long and more than five miles wide. An exploration discovery could potentially add trillions of cubic feet of gas to HUFFCO's proven reserves and millions of dollars to their bottom line. However, three uneconomic wells had already been drilled on this anticline. Although I could explain why the three wells were uneconomic, I knew this would be a high-risk exploration well.

HUFFCO's management and our partners approved the exploration well. To this day, I think the shortage of quality prospects strongly influenced the approval for this exploration well. My first exploration well with HUFFCO was a significant gas discovery. I breathed a huge sigh of relief when I knew we had a major commercial success.

Oil and gas exploration is a high-risk business, which requires significant capital to drill exploration wells. In 1980, there was only a 10% to 15% probability that an exploration well would be a commercial success. Multicompany partnerships were usually formed to mitigate the capital investment in high-risk exploration wells. After a commercial discovery, partnerships can become challenging if one of the companies in the partnership changes its corporate priorities or strategies.

Although I have usually found common ground with most partners, I will always remember one very challenging partner meeting in 1982. One of our partners had a representative that was notoriously prickly.

On a business trip to Houston, I stopped to see "Mr. Prickly" to discuss a new exploration idea I was developing. I went into detail on the new exploration idea, explaining the geological model, exploration risks, and significant oil and gas potential. I showed him several seismic lines to illustrate my exploration ideas.

At the end of the presentation, "Mr. Prickly" unlocked his drawer and slowly pulled out a single seismic line. The seismic line had been worn thin from the many changes in his interpretation of this single seismic line. "Mr. Prickly" then explained his exploration concept, which had many similarities to my exploration idea. The only difference between our two exploration ideas was the orientation, or a major bounding fault. However, the differences in our fault interpretation would not impact the location or depth of the well that I planned to propose at the next partner meeting.

"Mr. Prickly" asked me what I thought of his interpretation. I replied, "Your fault interpretation certainly could be right. I went with one structural model and you went with another. The key question is, do you think your company would support a well to test this new exploration play?" He said he would talk to his management, which meant: "No Comment."

One month later, we had a partner meeting in HUFFCO's Balikpapan office. In this meeting, I proposed a new exploration well, which I had previously discussed with "Mr. Prickly." In the presentation, I summarized the strengths and risks of the prospect and then asked for partner approval.

The first two partners immediately approved the well and said nice work. "Mr. Prickly," who represented the third company, said: "I think your geological and geophysical interpretation is completely wrong! However, my company's independent work shows a quality prospect, which can be tested by your proposed well location. Your proposed well depth is also acceptable, so my company approves HUFFCO's well proposal."

All the other partners turned their heads in shock at such a ridiculous and rude comment. I calmly said, "Thank you for the approval of this well." The exploration well was a significant discovery, which helped the partnership. As to "Mr. Prickly," well, he was always the prickly person in our partnership.

In my forty-year career, I have never worked with a more dedicated team of professionals. Management never demanded that we work long hours, six days a week. We all worked long hours because the work was exciting, and we also had to feed the eleven drilling rigs.

Saturday to me was a great way to catch up on my backlog of work. Saturdays also gave me some quiet time to think through challenging technical problems. One day, an Indonesian gentleman came into my office and introduced himself as Toto. He was the Chief Geophysicist for PERTAMINA in Kalimantan. His English was exceptional, and he wanted to discuss geophysical interpretation ideas and principles with anyone who had time.

Toto and I spent several hours discussing geophysical topics that were in the geophysical and geological journals that I received from the United States. Almost every Saturday, Toto would drop by my office to discuss geophysical articles and to discuss technical problems. I frequently lent Toto my technical journals and books. Little did I know that Toto and I would rekindle our friendship five years after I left Balikpapan, Indonesia.

Barbara quickly settled into camp living. She was active in a myriad of sporting activities. Shortly after we arrived in Balikpapan, she joined a group of expatriate wives on a boat trip, which traveled approximately one hundred miles up the Mahakam River. She had the opportunity to meet the warm and friendly indigenous Dayak people who lived along the river. Barbara was always up for an adventure.

Living in a camp, you learn to make do with what you have available. We held our own church services, theater productions, and sports competitions. We had many fun sporting competitions, but one sports activity, Hash House Harriers (HHH), provided many fond memories for both of us.

The HHH was founded by a group of British soldiers, before WWII. The soldiers would go to a Chinese eating house, known then as a hash house, outside of Kuala Lumpur, Malaysia. The soldiers would eat and drink beer. The soldiers would periodically get up from the table and run around outside the jungle paths to determine if they'd had too much to drink. If the soldiers had trouble making the jog, they would head back to the barracks.

Balikpapan had a men's HHH and a women's HHH running group. In 1980, HHH had evolved into an orienteering type of run. The people who set the run were called the "hares." The hares would lay a trail of biodegradable paper in the jungle. Our Balikpapan HHH runs would always start with an "off key" blast of a bugle. The bugle's bleat caused everyone to stampede down the trail with the shredded paper. The runs would always have several false trails, which were marked by the abrupt end of the paper on the trail. The lead runners would then have to call "check back," requiring the runners behind them to fan out and find the real paper trail. Once the real trail was found the runners would call "On-On," and off the group would go until the finish.

For some, HHH was a great form of exercise. The Balikpapan HHH fostered my passion for running. For others, the highlight of the night was the drinking and singing of "rowdy" rugby songs after each run. In Balikpapan, the HHH was the high point in the week for many of the expatriates.

In Balikpapan, the men's HHH was Monday evening and the women's HHH was Wednesday evening. The men's "Hash" had between seventy-five to one hundred and fifty expatriate and national runners. The women's Hash had a similar turnout. The only difference between the two groups was the women's Hash didn't have the excessive drinking or the "rowdy" rugby songs.

In 1980, Balikpapan's population was approximately two hundred thousand people. The town's existence was due solely to the exploration, production, and refining of oil and gas. The focus on the oil and gas industry in Balikpapan was similar to my experiences in Midland, Texas.

In 1980, HUFFCO, PETRONAS, Total (Total S.A.), and Union Oil were the largest oil companies in Balikpapan. HUFFCO and Union Oil were American companies, PETRONAS was the national oil company for the Republic of Indonesia, and Total was a French company.

HUFFCO and Total were in a very intense negotiation over the ownership of the giant gas field, Nilam. The giant gas field extended from HUFFCO's onshore license into Total's adjacent, offshore license. The two companies had to agree to a process to resolve the ownership of the Nilam field, precisely. Even a one percent change in the ownership in the field was worth tens of millions of dollars.

The negotiations over the process to resolve the precise ownership of the Nilam field dragged on for months. One Monday, we finally had a real breakthrough in our negotiations with Total S.A. Our general manager said, "Given our progress today, why don't I bring in dinner and we can finish our negotiation tonight?" The general manager for Total stood up and said, "But zee Hash," closed his briefcase, and walked out along with all his staff. The rest of us stood up and said: "But zee Hash," laughed with our general manager, and went to the men's Hash. Clearly, "zee Hash" was not an event to be missed for any reason.

Barbara and I had many wonderful Hash runs in Balikpapan. However, my most memorable run was set by a gentleman who was the head, or "grand master," of the Balikpapan Hash. The grand master was a Frenchman whose Hash name was "Captain Bagus" (Captain Good). Captain Bagus volunteered to set a run one night after perhaps partaking of a few too many beers.

Setting a Hash run through the jungles around the town of Balikpapan required extensive planning to avoid a disaster of a run. Anyone who set a poor Hash run always received enormous verbal abuse from their fellow members. This was to be expected, as the Balikpapan Hash was the event of the week.

Captain Bagus was a known procrastinator at his place of work at Total. As a true procrastinator, the good captain put off doing any planning for his Hash run. On the day of his run, Captain Bagus and one of his buddies left Total's office only two hours before the start of the men's Hash. The lack of planning predetermined a chaotic run of epic proportions.

Most Hash runs are a large loop, starting and ending at the same point. Captain Bagus's run was a point-to-point run. Taxis would take people from the finishing point of the run back to the place where the runner's cars would be parked.

With no preparation and insufficient time to lay a paper trail for even a basic Hash run, Captain Bagus came up with a panic-inspired plan. The good captain would lay a paper trail from the finish line toward the starting point, while his friend would lay a paper trail from the starting point toward the finish line. In Captain Bagus's muddled mind, the two men would meet somewhere in the middle of the run. Unfortunately for all involved, the two men both got lost and the two trails were never connected! Each man gave up and took a taxi back to the starting point of the evening's Hash.

Captain Bagus's flawed trail-setting technique delayed the start of the Hash by fifteen minutes. After much confusion, the Hash bugle was blown and off went more than one hundred runners into the dense jungles of East Kalimantan. Initially, the trail was wide, fast, and well-marked. After twenty minutes of hard running, the paper trail abruptly ended. All the runners fanned out to try and find the paper trail, which of course didn't exist.

I was running with a group of twenty men. A few men in our group thought we should run on a trail heading in a northwesterly direction. I pointed out that running northwest would be fine, if they wanted to run four hundred miles to Bintulu, Malaysia! However, I was going east, toward Balikpapan. Sanity prevailed, and our group headed east through dense jungle and swamp. By 8 p.m. it was dark, and we had to walk or else we ran the risk of stepping into a hole and breaking a leg or ankle. By 8:30 p.m. we stumbled into a small village and found out we were only a few miles from the main road. We left one runner with the village elder, since the young man was severely dehydrated.

My band of runners made it back to the "finish line" by 9:30 p.m. We found Captain Bagus lying semi-conscious by a fire drinking a beer. The cardinal rule in the Balikpapan Hash was the people who set the run are responsible for all the runners. The run organizers must send out search parties if anyone doesn't return to the finish line by 7 p.m. The only search party Captain Bagus had organized was the search for his next beer bottle!

After giving Captain Bagus a horrendous tongue-lashing, we made sure there were no more lost souls in the jungle of East Kalimantan. The next morning, I made sure the young man we left behind in the village was picked up and returned home. We also gave the village elder a few dollars for his hospitality and kindness, which he appreciated.

As to Captain Bagus, he was given a new Hash name, "Captain Tidak Bagus" (Captain No-Good). He was also verbally abused by the Balikpapan Hash for the next six months. We were all fortunate to survive this run, but it certainly provided me with a great story.

The Balikpapan Hash provided Barbara and me with many fond memories. However, one sporting event in Indonesia will be etched in my memory forever. Six months after we arrived in Balikpapan, the state oil company (PERTAMINA) approached all the oil companies and multi-national service companies about participating in a "friendly" competition. The invitation made it clear that all companies were expected to participate in the competition.

The competition included over thirty events, including soccer, basketball, volleyball, track, chess, and choral singing. I thought the primary goal was a friendly competition and to get to know our counterparts at the national oil company. Although that was the written goal, the primary and unspoken goal was for PERTAMINA to win the championship.

HUFFCO had seven expatriates that enjoyed playing basketball. At least five of our team members had played high school basketball and two had played college basketball. Our team easily beat the first three opponents by thirty or more points in each game.

Then we played PERTAMINA. Unlike the previous games, the basketball gym was packed with excited and noisy fans. There were over five thousand fans there to support PERTAMINA. After looking at the crowd, I asked Barbara to sit at the very top of the stadium just in case the game got out of hand. Little did I know what was to come!

Five minutes before tip-off, players from each team were told to line up single file. The referee instructed us to take off all jewelry, including my wedding ring. The referee then came along and clipped any unusually long fingernails! I realized this was going to be a very "memorable" event.

The HUFFCO team won the tipoff. When I got the ball, I passed it to one of our big men, who was mugged by three fist-throwing PERTAMINA basketball players. This set the tone of a very physical game throughout the evening. Our team displayed great control and never retaliated, which was wise since 99.9% of the fans in the gym were rooting for PERTAMINA.

Five minutes before halftime, I called a time out and went to the official's desk. I expressed my concern to the official that the game was getting out of hand. I asked that the referees please call a tighter game to reduce the number of hard fouls. I left the discussion knowing my words would have no effect on the referees.

When the game resumed, I drove to the baseline and was repeatedly fouled by a very aggressive PERTAMINA player. He then tried to push me out of bounds. As he was pushing me, I jumped up and passed the ball to one of our big men, who was standing near the edge of the court.

However, the basketball grazed the aggressive PERTAMINA player's leg. The ball went to our big man; however, the man guarding me proceeded to hit me with a right cross to my jaw. Fortunately, he grazed me, and the blow had no real impact. I remember thinking, "Keep cool—if you hit him you'll probably kill him, and the entire gym will break into a riot." As a result, I just turned my back and slowly walked a few steps away from him.

To my amazement, the referee actually called a foul on the PERTAMINA player. However, out of the corner of my eye, I saw a PERTAMINA fan jump out of his court-side seat and try to karate kick our big man, who was just holding the basketball, waiting on the referee. As the fan jumped out of the stands, his foot caught on the railing in front of his seat. The railing caused the young PERTAMINA fan to rotate in the air. The young fan then landed on his head and was completely knocked out! A hush fell over everyone in the gym.

After a few seconds, the entire gym broke into a riot! My first thought was to go to Barbara, who was at the top of the stadium seats. As I started to run toward my wife,

I heard whistles and saw more than fifty Indonesian police with shields and clubs moving into the gym. Fortunately, the police quickly brought order to the gym and no one was seriously hurt.

I knew the chief of police, so I went to him to apologize for this riot. He responded, "Oh, your team is fine; this is the first time we had to break up a riot at any of your games. We have had fights at all of PERTAMINA's basketball games." My first thought was: "So now you tell me!"

I thought the game organizers would surely cancel the HUFFCO-PERTAMINA basketball game. However, I had forgotten the unwritten rule: PERTAMINA must win the championship. The event organizers told HUFFCO that we would finish the game in no uncertain terms. We were then told the game would be played at 10 a.m. during a workday. We were also assured that there would be no fans attending the game.

We showed up at the designated time to find a gym packed with PERTAMINA fans. As captain, I looked at the other players on the team and said, "Our objective IS TO LOSE and get out of here alive!" We then lost and were given a hearty round of applause by the fans. This is when I learned the lesson of understanding the local culture and knowing the "real" objectives of any project you undertake.

In June 1982, I was given the opportunity to transfer to HUFFCO's headquarters in Houston, Texas. In my new position, I would work international new ventures, which was experience I thought I needed to gain to further my career. Barbara agreed with the move, as she saw this as an opportunity to further her education at the University of Houston.

As Barbara and I were preparing for our transfer to Houston, the United States was producing 8.65 Million Barrels of Oil per Day (MMBOPD), while consuming 15.30 MMBOPD a day. The average price for oil in the United States in 1982 was $31.83 per barrel. The average price for gasoline in the United States in 1979 was $1.22 per gallon, which is equivalent to $2.60 per gallon, when the price is adjusted to inflation (March 2015).

Rising oil and gasoline prices forced the world to conserve energy. In the United States, oil consumption steadily declined after 1978. Oil-importing countries also sought secure energy exporters. In Indonesia, major LNG export plants had recently been built in East Kalimantan and Sumatra. Indonesian LNG provided a lower export risk than oil from the Middle East. LNG also had a lower carbon footprint than coal or fuel oil. Japan and Korea became major importers of Indonesian LNG.

Energy conservation reduced the global demand for oil in the world. Indonesian LNG imports reduced the demand for oil in Japan and Korea. As the demand for oil decreased, major new oil discoveries were being made in the North Sea, West Africa, and Southeast Asia. The new oil discoveries would be producing oil into the world markets within five years.

The petroleum industry was turning a global oil shortage into an oil glut. In 1982, global energy data foretold of the coming oversupply of oil. As with any commodity, an oil glut would result in a dramatic drop in price. However, the economic "experts" were still projecting oil prices to be over one hundred dollars per barrel.

The rise in natural gas, oil, and gasoline prices spurred several green energy projects in the United States. In 1980, California funded the development of the first large-scale wind farm for electrical power generation in the world. This pilot project provided numerous environmental and engineering lessons, which were applied on future wind energy projects.

In March 1979, the Three Mile Island nuclear power plant in Dauphin County, Pennsylvania, had a partial meltdown. It is, to this day, the most significant nuclear power plant accident in the United States. The cleanup was completed in 1993 at an estimated cost of one billion dollars!

The Three Mile Island nuclear power plant accident galvanized antinuclear opponents across the United States. The Three Mile Island nuclear power plant accident contributed to the decline of new nuclear reactor construction in the 1980s and 1990s.

CHAPTER 7
Turkish Ice Cream

In 1982, the city of Houston, Texas, was growing at an unprecedented rate. Financial experts predicted oil prices would reach one hundred dollars per barrel within the next two to three years. In July 1982, the prime interest rate in the United States was 15.5%. Financial experts also predicted that the rising oil prices would increase the inflation rate in the United States economy.

If inflation continued to rise, the federal government would increase the prime interest rate. The financial experts' predictions created concerns that people would be priced out of the housing market. These concerns caused many people in Houston to buy homes beyond their financial means. In 1985, oil prices plunged, and oil companies began laying off tens of thousands of employees. People walked away from their homes because real estate prices plunged, and homeowners could no longer make their monthly loan payments.

We arrived in Houston in July 1982, and quickly found an apartment. Our apartment was almost as nice as our apartment in Balikpapan, Indonesia. Our new home was in the inner city, close to the University of Houston and my downtown office.

Barbara started university to gain her U.S. teaching credentials. In my new position, I would be working for two different business units, U.S. Drilling Fund and International New Ventures. I would soon be supporting a third business, which made life exciting, but very challenging.

In 1978, the U.S. federal government instituted an energy tax credit program. The program provided special tax credits for U.S. drilling funds. In 1982, HUFFCO was rated one of the top drilling funds. HUFFCO, as a result of their stellar financial performance, had little difficulty finding investors for their drilling programs.

The investors in HUFFCO's drilling fund were primarily insurance companies and pension funds. Our 1982 exploration-drilling budget was approximately one hundred twenty-five million dollars. During the year, we drilled a diverse range of exploration prospects, from offshore California to onshore Florida.

I found the investor selection process interesting, if not disconcerting. In the fall of the year, potential investors would send their financial managers to our office to review our drilling program for the next calendar year. All the potential investors employed a two-step decision-making process. Step one: an inexperienced MBA who had no knowledge of the oil and gas industry would receive a technical overview of our planned drilling program. Step two: a senior financial manager would receive an identical technical overview of our planned drilling program.

The senior financial manager usually had ten to fifteen years of investment experience, but only a rudimentary knowledge of the oil and gas industry. Within a few weeks, HUFFCO would receive confirmation of the investor's financial commitment for the upcoming year. In all cases, the due diligence by the financial managers was superficial at best. Even today, I have limited confidence in the oil and gas expertise of most financial experts.

In 1982, giant multinational companies like Chevron, Exxon, Gulf Oil, Mobil Oil, Royal Dutch Shell, and Texaco were spending billions of dollars a year on oil exploration in the U.S. Large multinational companies like Getty Oil, Occidental, Tenneco, and Superior Oil were spending hundreds of millions of dollars a year on oil exploration in the U.S. However, the real driving force behind oil exploration in the U.S. was the thousands of small, focused oil companies.

The small, focused oil companies ranged in size from just one geologist developing an exploration prospect to up to thirty technical professionals. These companies usually specialized in one region or basin in the United States. Most of the geologists in these companies knew the details of every well that had ever been drilled in their region. These professionals knew their region like the back of their hand.

The small, focused oil companies provided most of the exploration prospects to HUFFCO's drilling program. We would receive hundreds of drilling proposals every month from small companies all over the U.S. As a rule of thumb, I had to complete a thorough technical evaluation of thirty drilling proposals to find just one quality investment opportunity.

Supporting HUFFCO's drilling fund and the international new ventures business was a challenge. Every time I was called to work on an international new ventures project, my "in-box" continued to be filled with drilling fund proposals. I quickly found out that HUFFCO's drilling fund and the international new ventures business units weren't skilled at sharing resources, i.e. me.

HUFFCO's drilling fund unit employed approximately a dozen people in Houston and Corpus Christi, Texas. Evaluating the drilling fund proposals was interesting, since one prospect would be in the Rocky Mountains and the next would be offshore in the Gulf of Mexico.

However, my most vivid memories with HUFFCO's drilling fund are of some unscrupulous seismic contractors.

In the fall of 1982, oil prices began to drop, which rippled through the industry. The drop in the oil price meant a decline in corporate earnings. Lower earnings caused a swift reduction in the exploration work programs. Reduced work programs meant service companies shut down seismic crews and drilling rigs. Service companies are always the first to feel the impact of sliding oil prices.

One of my projects required the acquisition of new seismic data in onshore Louisiana. I initiated a tender to all the seismic contractors with seismic crews in the region. The sealed bids were submitted to me on a Tuesday by 4 p.m. Any company submitting a bid after the deadline was immediately disqualified from the tender.

I put all the sealed bids in a locked safe at the office and drove home. That evening, someone called me at home and asked to meet at the front of our apartment complex. The mysterious caller was a representative from one of the seismic companies that had submitted a bid earlier in the day. He told me his company really wanted this job and handed me an envelope, which contained several hundred dollars in cash. I immediately handed back the envelope and told him his company would now be disqualified from the tender. I also told him that if he ever tried this again, I would contact the police. He quickly left my apartment complex.

The next morning, I received a call from another seismic contractor that had also submitted a bid. The caller said his company would like to take me on an all-expenses-paid hunting trip to Wyoming. I commented that a trip like that would surely cost thousands of dollars. The caller laughed and said: "Absolutely!" I told the caller that if his company had taken the cost for the trip off their bid, they might have won the tender. I concluded by saying: "This is not how HUFFCO conducts a tender, and I find your approach unethical." I reviewed all the bids with our company's finance manager, and we selected the seismic contractor that had submitted the lowest bid and was technically qualified.

One of our exploration wells was a significant oil and gas discovery in Escambia County, Florida. The industry was abuzz with the news of this new exploration discovery and the oil and gas potential that could exist in this region.

A few weeks later, I received a phone call from one of the mineral lease owners near the exploration discovery. She asked if my company was acquiring new seismic data over the discovery. I replied no and asked what prompted her call. She told me there was a seismic crew working along the road outside her farm.

After a bit of detective work, I found that a well-known seismic contractor was acquiring "speculative" seismic data over HUFFCO's oil and gas discovery. Speculative seismic is data that is acquired by the seismic contractor and sold on the open market. The seismic contractor makes a fortune if he can sell the data multiple times but takes a loss if no one buys this data.

The seismic contractor was legally required to acquire HUFFCO's approval for this program, which they had failed to do. No company would approve this type of program since the company would lose their competitive exploration advantage in the region. I contacted HUFFCO's corporate attorney and then called the seismic contractor. I asked to speak to the vice president of speculative seismic operations. HUFFCO's attorney told the seismic contractor to immediately cease and desist their illegal operations. After a brief discussion, the vice president for the seismic contractor agreed to stop all operations in this region and agreed to pay damages to our company.

As we were concluding our discussions, I asked the vice president why he would take such a risk. He replied: "I rarely get caught, so it is worth the risk." I found this response unconscionable. However, I learned a valuable lesson to always keep up my guard in any future business dealings with all companies.

HUFFCO was always seeking new international opportunities. The international team consisted of a very small group of people with diverse nationalities and personalities. We worked very well together and developed an excellent camaraderie.

One of the geologists on the international team was required to fly to Balikpapan, Indonesia, for the first time. The geologist stopped by my office and asked me for guidance for his upcoming trip. I provided him with my insights and a very specific warning: "Do not eat any of the food from the stalls along the roads in Balikpapan." I explained that he would run a high risk of contracting hepatitis if he ate at one of those food stalls.

The international team always went out to lunch. Upon the geologist's return, I suggested we go to his favorite lunch spot. Much to my surprise, he declined. The next day the geologist showed up for work and his eyes were a pale yellow. Within one hour, a doctor was in our office giving us inoculations for hepatitis. I found out the geologist had ignored my warning and decided to try out several food stalls in Balikpapan and had contracted hepatitis.

After spending several months at home, the geologist returned to work. He was again asked to fly to Indonesia, but this time to fly to the capital city, Jakarta. The geologist again came to my office and asked me for guidance on his upcoming trip. I told the geologist that it was about a one-hour drive from the international airport to HUFFCO's Jakarta office. I then told him that taxi drivers in Indonesia are notoriously reckless drivers.

I then said with a straight face, "If the driver is going too fast, just say *'pergi cepat.'*" I didn't tell him that *"pergi cepat"* means "go fast." I then told him that if the driver didn't slow down, he should keep repeating *"CEPAT, CEPAT"* in a very loud voice. He proceeded to write everything down and left for the airport.

I was told the geologist's taxi ride between the international airport and HUFFCO's Jakarta office took less than thirty minutes. Both driver and the geologist were reported to have been sweating profusely. Of course, I received a good tongue-lashing from the geologist when he returned. However, after he calmed down, we both started laughing and remained good friends.

Our company's vice president of international operations asked me to help with our company's operation in Turkey. In 1981, HUFFCO was operating eight permits in southeastern Turkey. However, HUFFCO's Turkish operation had experienced three separate work-related fatalities on the seismic crews. One fatality was unacceptable, but three fatalities was unconscionable.

HUFFCO's Turkish operation had acquired two thousand miles of state-of-the-art seismic data at a cost of over four million dollars. The seismic data was sent to the Republic of Singapore to be processed. However, the final processed seismic sections looked like static on a television. Our vice president told me the Turkish Petroleum Ministry was ready to expel HUFFCO from the country because of the three separate fatalities on the seismic crews. Our two partners were ready to remove HUFFCO as operator due to spending millions of dollars on useless seismic data.

HUFFCO's Turkish operation suddenly became my highest priority. HUFFCO Turkey sent me the raw or unprocessed data for one seismic line for processing tests. I talked to a seismic processing company that was desperate for business. The manager of the seismic processing company agreed to do a pilot project at no cost to HUFFCO if I would work closely with his team. The seismic processing manager hoped his company would get seismic processing work from HUFFCO if he could demonstrate a dramatic improvement in the data quality. I knew this company very well, and I was confident in their capabilities. I also understood that HUFFCO's future in Turkey depended on delivering quality seismic data to our partners.

Over the next two weeks, I worked closely with the seismic processing team. I couldn't understand why the original seismic line, processed by the Singapore processing center, looked like static on a television. One morning, I got a telephone call from the seismic processing manager, who said, "I think you are going to like what you are going to see on the final-processed seismic line." I immediately rushed to his office and saw, to my delight, quality seismic data. I immediately went to our vice president's office, who also became very excited. Our vice president told me to telex the chief geophysicist in Turkey and to book a flight to Ankara immediately.

The next morning, I received a telex reply from HUFFCO's chief geophysicist in Turkey, demanding to know who gave me the authorization to reprocess the seismic line. He must have forgotten that he had sent me the raw or unprocessed seismic data from Turkey. As I was staring in disbelief at the telex, our vice president came into my office and told me not to worry about the telex, but to fly to Ankara, Turkey, as fast as possible.

Our vice president then went to his office, closed the door and called the general manager and the chief geophysicist in Ankara, Turkey. Given our vice president's booming voice, it didn't matter that his office door was closed. The phone call from our vice president was colorful and direct. I was to be given all the support I needed to try and right the geophysical operation.

My flight to Ankara, Turkey, would take me on Pan American Airlines from Houston, Texas, to JFK International Airport in New York City and then to Frankfurt, Germany. I would then fly on Lufthansa Airlines from Frankfurt, Germany, to Ankara, Turkey. This was my first flight on Lufthansa and it would prove to be memorable.

In the 1980s, airlines had smoking and no smoking sections on the plane. As a nonsmoker, I always reserved a seat in the nonsmoking section of the plane. My airplane to Ankara had two rows with two seats in each row. As I took my seat, I checked with the flight attendant to be sure I was sitting in the nonsmoking section of the airplane.

As soon as the plane took off, the gentleman in the seat in the aisle seat opposite me lit up a Turkish cigarette. Turkish cigarettes have a strong and very pungent odor. I immediately asked the flight attendant if she would ask the gentleman to stop smoking, since we were in the nonsmoking section.

To my astonishment, the flight attendant told me the smoking section was on the right side of the plane and the nonsmoking section was on the left side of the plane. It seems the Lufthansa flight attendants had lobbied for this innovative idea, which left me speechless. As a result, I had an intense, four-hour secondhand smoke experience, courtesy of Lufthansa Airlines.

Once in Ankara, I immediately went by taxi to my company's office. Although I had a cool reception from the chief geophysicist, the general manager went out of his way to help me settle into my office. I had previously worked with all three of the geologists in the Ankara office. The geologists couldn't understand how the Singapore processing center had failed to deliver useable seismic data. However, everyone was excited about the seismic processing results from the processing center that I used in Houston. I spent the next four weeks getting a crash course in the geology of the area and prioritizing which seismic lines should be reprocessed by the Houston processing center.

Over the next six months, I flew from Houston to Ankara every other month. Once in Ankara, I would spend four to five weeks working with the geologists to develop exploration prospects. After each trip to Ankara, I flew back to Houston to catch up on a mountain of submittals that were stacked on my desk from the company's drilling fund.

On my first trip to Turkey, I bought a book to learn how to speak a few words of the country's tongue twisting language. My goal was to be able to order lunch and to speak a few words to HUFFCO's Turkish employees. At lunch, I always went to a small café near the office. The two waiters in the café didn't speak English, but the menu had pictures of the main dishes. I just pointed to a picture on the menu and within a few minutes a delicious pizza arrived. Once I finished lunch, I would make a writing sign in the air with my hand, which resulted in the bill.

After studying my English-Turkish book, I thought I would try out my newfound "language skills" at my local café. It was a very cold, snowy day in November when I went for lunch. As usual, I pointed to the pizza of my choice on the menu. Once I finished my pizza, I thought I said in Turkish, "May I have the bill, please?" My words immediately drew a shocked look from the faces of the two waiters. However, one waiter held up his index finger and immediately ran out the door. Thirty minutes later, he returned with a bowl of ice cream. I am sure I also looked surprised when the waiter put down the bowl of ice cream in front of me. However, I just smiled, nodded, and proceeded to eat the ice cream. I then returned to hand signals to request the bill for my lunch. I learned a valuable lesson—know your limitations!

The safety issues on the seismic crews were addressed, and the Turkish Petroleum Ministry allowed HUFFCO to continue with their planned-exploration program. The seismic reprocessing in Houston delivered excellent results. We were now developing an inventory of quality exploration prospects.

On my fourth trip to Ankara, I began preparing for the next partner meeting. It was very important to demonstrate to the two partners that HUFFCO should remain operator of this operation. Although the partners were pleased with the recent reprocessing results, they expected to see an inventory of quality exploration prospects. I met with the drafting supervisor and gave him a ranked list of work that needed to be completed for the partner meeting. I divided the drafting work into three tiers, based on importance. Tier 1 work had to be completed by the time I returned to Ankara in another four weeks. Tier 2 work was not critical, but it would be nice to have for the partner meeting. Tier 3 work was low priority and could be completed whenever he and his staff had time.

The drafting supervisor had served in the Turkish Army and was disciplined and reliable. He smiled and told me not to worry; the work would be done. He also assured me that he knew how important the upcoming partner meeting was for HUFFCO's future in Turkey.

On my fifth and final trip to Ankara, I took a taxi directly to the office to prepare for the partner meeting, which begin in two days. I worked in my office until 10 p.m. before I left my office to check into my hotel. As I was leaving my office, I noticed the lights in the drafting room. When I went into the drafting room, all five draftsmen were working. It was obvious they had been working a very long day.

I asked the drafting supervisor the status of the work I had given him. He told me the last item on the Tier 3 list would be completed that night. I immediately apologized and said, "I must have misspoken, since the Tier 3 drafting work was to be done only when you had nothing else to do." The drafting supervisor said: "I understood what you said, but every time you are here, you work twelve or more hours a day, seven days a week. You are one of the few expatriates who are doing everything they can to help the company succeed. Our jobs depend on the success of the company. We all want to make sure you and the company are successful." His words made my many long days and night worthwhile. This is my fondest memory of my time in the Republic of Turkey.

Our partner meeting went very well, and all the companies agreed to drill several exploration prospects. More importantly, the partners regained confidence in HUFFCO as the operator. In our partner meetings, I developed a friendship with a geologist with one of our partners. His company had seconded him to HUFFCO's Ankara office to gain operating experience. We would renew our friendship in another three years in Jakarta, Indonesia.

Once the work in Turkey was completed, I spent all my time on the company's drilling fund and international new ventures. The owner of the company was also a geologist. The owner provided our new ventures team with explicit criteria. The owner's criteria allowed our small team to efficiently evaluate opportunities all over the world.

Our international new ventures team ranked all the countries in the world on hydrocarbon potential and political risk. Ranking a country's hydrocarbon potential required an analytical analysis of existing oil and gas fields and a conceptual assessment of new geological ideas or plays. The political risk evaluated the frequency and severity of changes the government made to existing hydrocarbon contracts.

Our analysis showed that a country like Australia had one of the highest political risks in the world. The Australian government had frequently changed existing contract terms over the previous ten years. In one case, the stroke of a bureaucrat's pen turned an economic oil field into an uneconomic oil field. The changes to the tax codes resulted in the loss of hundreds of jobs and millions of dollars in revenue to the Australian government. Countries such as Malaysia and Indonesia had very low political risk, since these governments understood that businesses would invest in countries that have stable fiscal and tax policies.

Ecuador was one of a handful of countries that our new ventures team rated as having high oil potential and low political risk. The national oil company of Ecuador, Corporación Estatal Petrolera Ecuatoriana (CEPE), had announced plans to hold a major hydrocarbon licensing round. In the spring of 1983, the senior geologist on our new ventures team and I flew to Quito, Ecuador, to meet CEPE to discuss the upcoming hydrocarbon licensing round.

Our flight took us from Houston, Texas, to the Miami International Airport and then to Quito, Ecuador. The only disconcerting aspect of the flight was seeing several crashed planes in the mountains shortly before we began our descent into Quito, Ecuador.

As soon as we checked into the hotel, I went for a run to help me recover from the long flight. I ran for what I thought was one hour. When I checked my watch, I saw that I had only been running for thirty minutes. I then realized the difference between running at sea level in Houston and at more than nine thousand feet in Quito, Ecuador.

The geologist and I decided to walk around the town to find a restaurant for our evening meal. We walked into one vacant restaurant after another. We couldn't understand why no one was in any of the restaurants. Finally, we heard loud cheers and shouting in the very back of one restaurant. We went into the kitchen and found all the staff listening on a radio to a soccer match between Argentina and Ecuador. Argentina was a world powerhouse in soccer. However, Ecuador was leading Argentina 1-0 in the game. In the last few minutes of the game Argentina scored the equalizing goal. However, everyone started celebrating when the game ended in a 1-1 draw. Ecuador had been such an underdog in the game that a draw was a victory to these passionate soccer fans.

We had a wonderful supper at the restaurant and enjoyed watching people celebrate the outcome of the soccer game. The next morning, we met with the CEPE team that would be coordinating the hydrocarbon licensing round. In the meetings, we found out the hydrocarbon licensing process, license award selection criteria, and bid submission dates.

In this meeting, I met Carlos, the chief geophysicist for CEPE. He was a very knowledgeable geophysicist with whom I quickly established a rapport. Carlos was very interested in my experiences with jungle-portable seismic crews in Indonesia. During our discussion, I asked him why one of the blocks in the upcoming licensing round had absolutely no seismic data over the western half of the block.

Carlos leaned over and asked me, "What do you think of analog seismic data?" I replied, "I worked on an analog seismic crew in Texas and West Virginia. Analog seismic data isn't as good as modern, digital seismic data. However, I would welcome the opportunity to see any analog seismic data over the blocks in the upcoming licensing rounds." Carlos led me to a room filled with paper analog seismic data, which covered the western half of the block with no seismic data.

It seems another company had visited with Carlos prior to our arrival. Carlos asked this company if they would like to view the vintage analog seismic data. The geophysicists from the other company apparently laughed at Carlos for suggesting they spend time looking at such old and useless seismic data. This would prove to be a lost opportunity for more than one of the companies in the license round.

The data quality of the analog seismic data was very good. Within a few hours, I mapped several exploration prospects in the western half of what I thought was the most promising block. With Carlos's permission, I took my maps back to Houston.

Our senior geologist became the team leader for the Ecuador evaluation. He and I spent many late nights looking at thousands of miles of seismic data, hundreds of reports, and dozens of well logs. In August 1983, we were on schedule to complete a rigorous evaluation of all the blocks in the upcoming licensing round. However, we had not accounted for Mother Nature, specifically hurricane season!

In August 1983, Hurricane Alicia was growing in intensity in the Gulf of Mexico.

Alicia grew into a Category 3 hurricane and slammed into Galveston, Texas, with wind gusts of well over one hundred miles per hour. Alicia then headed directly toward Houston, losing little of the storm's intensity. Hurricane Alicia passed directly over the city center of Houston, as shown in *Figure 9*.[14] Houston received over eleven inches of rain within a few hours, causing massive flooding. The hurricane force winds knocked out the power lines throughout the city.

Figure 9

Hurricane Alicia caused twenty-one deaths, thousands of injuries, and billions of dollars in damages to the Houston area. To the city's credit, the power companies initiated a swift and effective plan to quickly restore power. The police and fire departments also did a phenomenal job of helping distressed families and maintaining order during this time of crisis.

Barbara and I escaped Hurricane Alicia unscathed. We were also fortunate since our apartment suffered no significant damage. However, it took the power company several weeks to restore the electricity in our apartment. Hurricane Alicia gave me a new appreciation for air conditioning, since the summer temperatures in Houston routinely exceeded 90 degrees Fahrenheit with over 90% humidity.

The next morning, I drove downtown to HUFFCO's office building to assess the damage. I kept thinking about how we were going to complete the Ecuador license evaluation in time to submit a bid. The hurricane winds had created a twisting motion in the skyscrapers. The twisting motion caused the large glass windows in almost

---

14   NOAA National Climate Data Center

every office building to pop out and crash on the sidewalk below.

I talked to the fire marshal, and he said we could return to work the following day if the building manager could get the window glass replaced or at least covered with plywood.

The building manager couldn't get the glass replaced, but he did get large sheets of plywood nailed over all the windows. The electrical power was also working so we could continue our technical evaluation. Unfortunately, the wood over the windows didn't seal, so the humidity inside of our office building was over 90%. Everyone was wearing shorts and flip-flops during the evaluation. We completed the technical evaluation on time, thanks to the help from Houston civil servants, and of course, very long hours of work.

We presented our technical and economic evaluation of Ecuador to HUFFCO's owner. He told us this was just the type of opportunity he was seeking; however, financially he could not proceed with the project. Unfortunately, HUFFCO had purchased a drilling company and had built a heavy oil refinery in California. Although Indonesia was generating significant cash flow for the company, the drilling company and the heavy oil refinery were losing significant money.

HUFFCO had two companies join the Ecuador evaluation. One of our partners still wanted to participate in this license round. This partner asked our owner if he would find another company for them to join in the upcoming license round.

Our Ecuador team leader knew that Conoco, a large, multinational oil company was seeking another partner. He put the two companies in contact with one another. A few days later, we were asked to give our technical presentation to Conoco. I thought this was an unusual request, as large, multinational oil companies usually have numerous technical professionals working on these types of licensing rounds.

Conoco's technical team and their new partner listened, while our team leader addressed the regional geology of the Oriente Basin in Ecuador. I presented the geophysical interpretation and prospect assessment. Conoco's chief geophysicist was surprised to learn of the existence of the vintage analog data covering the western area of one block. The chief geophysicist was also surprised over the rigor we had taken to ensure accurate positioning of all the offset wells and seismic lines. At the end of my presentation, the chief geophysicist asked me how many geophysicists we had working on this project. He was very surprised to learn that I was the only geophysicist on HUFFCO's evaluation team.

One week later, I received a call from Conoco's chief geophysicist. After a bit of small talk, he asked me if I would be interested in a position with Conoco. The chief geophysicist was apparently impressed that I had turned out more maps than his team of six geophysicists. He said he would like me to come in for a job interview. I told him I first had to talk to my wife, but I would call him back the next day.

That night, Barbara and I discussed the possibility of me leaving HUFFCO to return to work for a large, multinational oil company, like Conoco. Although I enjoyed working for HUFFCO, it was obvious the company was going to continue to have financial problems. Those problems would become serious, if the oil and gas prices continued to fall.

I had thought about starting my own geophysical consulting company. I had received several calls from companies that wanted me to work for them on domestic and international projects. However, Barbara and I enjoyed living overseas. We found overseas living gave us the opportunity to experience different cultures and travel the world. As a consultant, we would almost certainly spend the rest of my working life in Houston.

The next morning, I scheduled my job interview with Conoco. The interview went well, and I was impressed with the professionalism of all the people I met. The following week, Conoco called with a job offer to work in London, England, with an international "new ventures" team. After discussing our options with Barbara, I accepted Conoco's offer.

As Barbara and I were preparing for our move to London, the United States was producing 8.69 Million Barrels of Oil per Day (MMBOPD), while consuming 15.23 MMBOPD a day. The average price for oil in the United States in 1983 was $29.08 per barrel. The average price for gasoline in the United States in 1983 was $1.16 per gallon, which is equivalent to $2.37 per gallon, when the price is adjusted to inflation (March 2015).

High oil and gas prices had fueled a surge in global exploration. The search for new reserves resulted in numerous, giant oil discoveries in South America, West Africa, North Sea, and Southeast Asia. The oil industry was rapidly developing the major new oil discoveries. Oil from some of these new discoveries was just beginning to reach the global market. The private sector's search for new oil reserves was rapidly changing the global supply and demand dynamics for oil.

CHAPTER 8

Bombs and Strikes in London

Barbara and I arrived in London in November 1983. We were delighted to be abroad, especially in a country with such a rich and storied history. We also anticipated an easy adjustment into the British culture. However, I soon learned that American and British cultural differences were much greater than I'd imagined.

During the 1970s, Great Britain's economy became stagnate, the result of unwise government policies, economic crises, and prolonged labor union strikes. In 1979, the economic stagnation led to the election of the Conservative Party's Margaret Thatcher as prime minister in 1979. The "Iron Lady," as the prime minister was called, set out a controversial program of privatizing major segments of the British economy, including the coal, iron, steel, natural gas, telecommunication, and transportation industries.

Resistance to Prime Minister Thatcher's program quickly turned violent. London felt like a city under siege. There was a significant police presence throughout the city due to the threats from the Irish Republican Army (IRA). Britain's refusal to remove military and police peacekeeping forces from Northern Ireland led to an IRA declaration of "economic war" on England, which included bombing commercial targets in London.

In January 1984, I was working in my office on Sunday when I heard a muffled bang. Hearing no other noise, I went back to work. I finished an hour or so later and left the office through the building's back exit. As I started to walk home, three British police (or "bobbies") accosted me, demanding to know who I was and what I was doing in the area. I produced my passport and the bobbies let me continue my walk home. The next morning, I read in the newspaper the IRA had detonated a bomb in front of a store just a few blocks from our office. Fortunately, all the stores were closed on Sunday, so no one was injured.

In addition to problems with the IRA, the prime minister was preparing for a major test of wills with the National Union of Mineworkers (NUM). Mrs. Thatcher wanted to close unprofitable coal mines and privatize the rest of them. NUM President Arthur Scargill was vehemently opposed to the prime minister's plans. Tension between these powerful adversaries was palpable and could be felt across Britain. The British Broadcasting Corporation (BBC) provided almost daily updates on the escalating tension. The tensions from the IRA bombings and the unannounced bus and train strikes in support of the NUM took a toll on everyone living and working in London.

Barbara and I were blissfully unaware of the crisis atmosphere in London when we landed at Heathrow Airport. We took a taxi to a very nice hotel just a few blocks from Conoco's office. My first morning in London, I bought several newspapers to read at breakfast. I was taken back by the tabloid style of journalism in the well-known and well-respected papers.

Our hotel had placed promotional brochures in our room touting its "famous English breakfast." Unfortunately, the breakfast consisted of vegetable and gristle sausages, cold runny eggs, and the worst coffee I ever tasted. This was my first of many lessons in the cultural differences between Britain and America.

After a few days in the hotel, we set out to find an apartment ("flat" in the local dialect). This task took nearly two months. We rejected many flats for significant structural faults, broken water heaters, major plumbing problems, leaking rooves, etc. Many of my colleagues elected to rent cottages forty to fifty miles outside London in the picturesque countryside. However, my colleagues who became country gentlemen were faced with daily commutes of more than four hours.

After extraordinary effort, Barbara found a flat in an old home in Hampstead Heath, close to public transportation and an easy walk to the shops. It took me only twenty minutes to get to work by bus. Our flat was spacious, though drafty and poorly heated, but we had wonderful neighbors and access to a lovely back garden. When the public transportation labor union called a strike, I simply jogged home after work while my "country gentlemen" colleagues either slept on their office floor or paid a king's ransom for a room in a London hotel.

Conoco's recent new ventures work in the Netherlands offshore resulted in two commercial oil fields. The company opened an office in Leidschendam, Netherlands, to handle the field development and production operations. Our new ventures team in London then moved to the next area of interest, West Africa.

Conoco had acquired a nonoperating interest in an exploration permit in offshore Gabon. The operator of this permit was an American independent company that had drilled one well in an exploration prospect. Although the results of the first well were promising, follow-up drilling condemned the economic potential of this prospect.

Our London office was given the responsibility of working with the operator to try and salvage some value from the offshore Gabon permit. The operator was very aggressive and proposed the partnership drill four more exploration wells in the permit. A surprising proposal, as the operator had identified zero prospects in the permit. It was clear that the operator was desperate. It would be another year before I found out the reason for the operator's desperation.

Our team identified one relatively low-risk exploration prospect. After additional technical meetings, the partnership agreed to drill one more exploration well. Fortunately, the exploration prospect turned out to be a commercial oil discovery. The following year, the American operator sold off all its global oil and gas assets. The asset sale was necessary to pay down the debt of the parent company that was nearly insolvent.

I was asked to lead the evaluation for an upcoming offshore licensing round in Gabon. This would be my first encounter with an African government. I spent almost all my waking hours evaluating the oil and gas potential of Gabon. In 1984, Gabon was the third largest hydrocarbon producer in Africa. Two large multinational companies, Elf Aquitaine and Royal Dutch Shell, dominated the Gabonese petroleum industry. Limited industry competition indicated to me that Gabon had significant undiscovered oil and gas potential.

All too many countries in West Africa have suffered from tribal wars, plague, famine, coup d'états, high crime rates, and rampant corruption. My research on Gabon indicated the country had experienced almost none of the misery that kept other West African countries in abject poverty.

Gabon had achieved more than fifteen years without tribal war, coup d'états, or violent crime. The principle reason for Gabon's peace and modest prosperity was the president, Omar Bongo. The president was not a saint, but he was a crafty politician. The president coerced world health and aid organizations into giving assistance to his country, even when other West African countries had far greater needs. The president converted to Islam to gain Gabon access to OPEC. OPEC also gained Gabon additional financial aid from several wealthy countries in the Middle East.

President Bongo placed relatives in key cabinet positions and systematically paid off all rivals to ensure he would not have any opposition. The president also forged close economic and military alliances with France. In 1984, the largest continent of French Legionnaires was in Gabon.

Although President Bongo maintained peace, he squandered billions of dollars of oil revenue on his personal comforts and patronage. Education and training was never a priority for the Gabonese government. President Bongo believed it was easier to control an uneducated country.

Our team completed the initial geological assessment of Gabon in London. We were now ready to fly to Libreville to meet with the government officials that would be handling the upcoming offshore license round. We took a direct flight from the London Gatwick Airport to Libreville, Gabon. We arrived late at night and, as we disembarked, I noticed dozens of heavily armed soldiers throughout the airport. I was told the airport always maintained a high level of security, which was comforting to a degree.

We stayed at one of several five-star hotels in Libreville. The food at the hotel was exceptional. I found it surprising that the hotel manager and his direct reports were all French nationals. The next morning our team arrived at the Ministry of Mines, Petroleum and Hydrocarbons, for our scheduled meeting. I was shocked to find that many of the bureaucrats at the Ministry of Mines, Petroleum and Hydrocarbons, were French nationals. Although all the government officials were courteous, it was clear that the French government planned on maintaining a dominant presence in Gabon.

In most countries, training and development of national staff is considered a critical criterion in the license round selection process. After our meeting with the Ministry of Mines, Petroleum and Hydrocarbons, I met with department heads at Omar Bongo University in Libreville. As I expected, all the department heads in science and engineering at the university were French nationals. In my trip report, I described the "Gordian knot" the French government had put in place to thwart the successful entry of non-French companies, like Conoco, into Gabon.

More than a dozen licenses would be available at the upcoming Gabon licensing round. Upon our return to London, our team began a thorough analysis of each of the offshore licenses. In only four weeks, our team evaluated thousands of miles of seismic data and hundreds of well logs in Gabon. After four intense weeks of work, our team identified only one prospective license. Our next step was to fly to Houston, Texas, to present our technical evaluation and recommendation to Conoco's senior management.

The geologist and I made the technical presentation to the vice president of international exploration and his management team. I was impressed with our vice president since he was very knowledgeable about the geology of Gabon and all the relevant technical issues. I made the recommendation to bid on the prospective lease. I then summarized the strengths and technical risks of the recommended off-shore Gabon license.

The vice president abruptly stopped the meeting. You could hear a pin drop in the meeting room when he said: "This is the first time anyone has ever summarized the risks, as well as the strengths, of an opportunity. It is time we quit selling and start making fact-based recommendations." The vice president approved our recommendation to bid a multimillion-dollar work program for the offshore Gabon license. The next morning, we flew back to London and began preparation for the bid submission to the Gabonese government.

Barbara and I decided to take a vacation after the Gabon offshore bid submission was completed. We decided to take a cycling vacation in the Jura region in eastern France. The cycling company, "Cycling for Softies," provided bikes, route map, hotel or pension reservations, and emergency contacts. There was no support vehicle and we had to carry all our clothes and basic necessities on our bicycles for the ten-day trip.

Typically, we would cycle twenty-five to thirty miles from one village to the next. We would then spend one day in the village, cycling around the back roads and viewing the beautiful countryside. We would then repeat the process until we returned to the starting point of our trip.

Barbara spoke a little French, which helped when we occasionally got lost. One day after a long ride, Barbara decided she wanted to take a bubble bath. She asked me to go to the small grocery store opposite our pension and buy bubble bath for her. Given my linguistic experiences in Turkey, I had very little confidence that I would be successful in finding bubble bath.

Barbara gave me a small English to French dictionary, as I didn't speak any French. Barbara thought nothing could go wrong with such a simple task. Unfortunately, the dictionary didn't contain the word "bubble bath" and no one in the small grocery store spoke English.

What transpired can best be described as an impromptu game of charades. I would make gestures with my hands or point to words in the dictionary, and the crowd in the store would shout words enthusiastically in French. One gentleman threw up both hands, grabbed my arm, and pulled me to a row of bins. He pulled open one bin and said "Voilà!" as he pointed to a container filled with dried pinto beans. At least I didn't get ice cream, like I did in Ankara, Turkey!

I was finally able to locate bubble bath, thanks to the patience of the storeowner and the enthusiastic townspeople. The people we met in the Jura region were always friendly and we had a wonderful cycling trip.

Upon my return to London, I learned that Conoco had been awarded the Gabonese license. Although I was confident of our technical assessment, I knew winning the license wouldn't guarantee success. Historically, exploration projects in offshore Gabon only had a 25 % probability of finding commercial quantities of oil.

Conoco was operator with a 100% interest in the offshore Gabon license. Our team was asked to find a partner to mitigate the exploration risk. In relatively short order, we located one large multinational company to join Conoco in the Gabon exploration program.

After the Gabon license round, I was asked to evaluate the oil and gas potential for offshore eastern Greenland. No wells had been drilled in offshore eastern Greenland. The only available data to complete this evaluation were a few, very long regional seismic lines. My evaluation had enormous uncertainty due to the very limited data.

Offshore eastern Greenland offered only one certainty: extreme arctic weather. Icebergs routinely drifted from the northern area of the island to the southern area of the island. The cost to build offshore platforms that could withstand the extreme arctic weather exceeded the value of the oil potential for the region. My presentation to Conoco's management quickly ended any further interest in offshore Greenland.

After completing the Greenland evaluation, I was offered a promotion and a transfer to Conoco's Indonesian operation. Over the previous decade, Conoco had drilled over one hundred consecutive, uneconomic exploration wells in Indonesia. My previous company, HUFFCO, was one of the most successful exploration companies in Indonesia. I was now working for the most unsuccessful exploration company in Indonesia. I think Conoco's vice president of international exploration thought I might bring a fresh perspective to our Indonesian operation. I might bring a fresh perspective, but would anyone listen to someone who was new to Conoco?

As Barbara and I were preparing to leave for Jakarta, Indonesia, the United States was producing 8.91 Million Barrels of Oil per Day (MMBOPD), while consuming 15.7 MMBOPD a day. The average price for oil in the United States in 1985 was $26.92 per barrel. The average price for gasoline in the United States in 1985 was $1.12 per gallon, which is equivalent to $2.14 per gallon, when the price is adjusted to inflation (March 2015).

In 1985, the United States federal government allowed the drilling tax credit policy to expire. The expiration of the drilling tax credit policy resulted in over ten billion dollars being withdrawn from drilling funds in the United States. The U.S. government recognized that the global supply of oil was catching up with the demand for oil. As a result, the drilling tax credit policy would no longer be necessary for the United States economy.

From 1960 to 1980, the U.S. nuclear power industry experienced a revolution in technology and growth. However, the shining star of nuclear power quickly faded due to massive cost overruns. Existing power plants required expensive modifications due to new government safety regulations. Construction of new nuclear power plants were delayed due to continuous changes in nuclear safety regulations and opposition from antinuclear energy organizations.

In 1985, an article in *Forbes Magazine* stated: "The failure of the U.S. nuclear power program ranks as the largest managerial disaster in business history, a disaster on a monumental scale . . . only the blind, or the biased, can now think that the money has been well spent. It is a defeat for the U.S. consumer and for the competitiveness of U.S. industry, for the utilities that undertook the program and for the private enterprise system that made it possible."[15]

The escalating cost for electricity from nuclear power plants made fossil fuels far more cost-effective for utilities. However, the supply and demand for fossil fuels would prove to be cyclic, creating dramatic price volatility over the next decades. Once again, the United States was faced with an energy conundrum of finding sustainable, reliable, low cost energy for electricity and for transportation.

---

15   "Nuclear Follies," James Cook, *Forbes Magazine*, February 11, 1985

CHAPTER 9
Turning the Tide in Indonesia

Barbara and I couldn't move to Indonesia until I received a provisional work permit from the government agency, Badan Koordinasi Kontraktor Asing (BKKA). The work permit process required me to submit a detailed education and work history to the Indonesian Embassy in London. Conoco's Jakarta office told me it usually took three to four months to gain a "provisional" work permit and another two to three months for the final work permit.

Technical professionals (geophysicists, geologists, reservoir engineers, etc.) that are posted overseas in developing countries, such as Indonesia, are expected to train and develop the national technical staff. In 1985, prior to the internet, technical professionals usually carried reference books to train the national technical staff. The time it took to gain an Indonesian work permit was fast, compared to the government's customs agency. I knew Indonesian customs could take six to twelve months to "process" my technical books into the country.

Conoco provided me the opportunity to visit the Jakarta office while my work permit was being processed. I saw my visit as a way to expedite the "process" of shipping my technical books to Indonesia. For my Jakarta trip, I packed one small suitcase with clothes and two boxes of technical books, which weighed more than fifty pounds!

As soon as I landed at the Jakarta airport, I was pulled aside by a senior customs officer, who wanted to know what I had in the boxes. I immediately smiled and said "Selamat pagi tuan." (Good morning, sir) in my fractured Bahasa Indonesian accent.

The senior customs officer said in flawless English: "You speak Bahasa!" I explained in English that I had learned a few words of Bahasa when I first lived in his beautiful country from 1980 to 1982. We then began to discuss my experiences living in Borneo, his family, and the challenges of his job as a senior customs officer.

Standing a few feet behind me was a Japanese businessman who clearly was in a hurry. It was evident that the businessman didn't appreciate the friendly conversation I was having with the senior customs officer.

My conversation with the senior customs agent slowly moved to the topic of my fifty pounds of technical books. I explained that, as Conoco's chief geophysicist, one of my primary responsibilities was to train and develop young national professionals who could then replace me within the next few years.

The senior customs officer said, "We will just keep your books at the airport and you can drive to the airport anytime you want to train the young national professionals." I smiled and said, "That is a wonderful idea; however, it is a long drive back and forth from our office to the airport. Perhaps I could pay a tax that would allow me to keep the books at our office?" After negotiating back and forth for another fifteen minutes we agreed on a price for the tax, which was the equivalent of two U.S. dollars for more than fifty pounds of books.

I could tell the Japanese businessman was becoming increasingly irritated, as he was pacing back and forth. The businessman's actions were also noted by the senior customs officer. I concluded my conversation with the senior customs officer by thanking him for his time and wishing him and his wonderful family well.

As I was putting my two heavy boxes of technical books on a cart, I saw the Japanese businessman open his small briefcase, which contained only three small books. The Japanese businessman asked the senior customs officer curtly: "How much?" The businessman then opened his wallet and pulled out a crisp one hundred dollar bill. The senior customs office then turned to me, smiled, and said back to the Japanese businessman: "It will be that one hundred dollar bill and nine more of his brothers."

After I cleared customs, I took a taxi to Conoco's office to meet the exploration manager, who was my new boss. My boss and I had a pleasant conversation. However, I could tell he was already feeling pressure to turn around Conoco's abysmal Indonesian exploration program.

I only spent a few days in Jakarta before I flew home to London. When I arrived at the London office, I had a note to call my new boss in Indonesia as soon as possible. When I called my boss in Jakarta, his very first words to me were: "Who do you know at BKKA?"

It seems my final work permit had been approved by BKKA in only three weeks. My new boss was concerned that I had bribed a government official to expedite the approval of my work permit. I told my new boss I didn't know who approved my work permit. Once I arrived in Jakarta, I found out that the chief geophysicist for BKKA was my old friend, Toto, who I knew from my days in Balikpapan, Indonesia.

Barbara and I started packing for Indonesia once I had my final work permit. The only challenge with our move was the importation of Barbara's pet cat, "Snookiepuss," from England to Indonesia. We had acquired Snookiepuss at a humane society in the United States and had brought her with us to London. We had tomes of medical reports on Snookiepuss, as the British are very strict when it comes to importing pets into their country. We also checked with the Indonesian Embassy to make sure we had all the necessary paperwork to import Snookiepuss into the country.

Barbara, Snookiepuss, and I flew from Heathrow Airport to the international airport in Jakarta. We had a turbulent flight, and the seat belt sign never went off through the entire flight. When we landed, we picked up our bags, including the cat-carrying cage. When I picked up the cat carrier, it came apart, as the screws in the plastic shell had vibrated off during the turbulent flight. I quickly reassembled the cat-carrying cage, complete with one very scared Snookiepuss.

As soon as we got in the customs line, we were approached by a customs officer who took me to a room in the bowels of the airport. He looked over the British paperwork for the cat and said, "These papers are not in order." When I asked what was needed to gain Snookiepuss admission into Indonesia, he smiled and said the equivalent of twenty-seven U.S. dollars. I paid the money and then went back to the customs line with my wife, cat, and luggage.

Just as we were getting to the front of the line, another customs agent approached and pulled me out of line. Unfortunately, he was not the same friendly customs officer I got to know on my initial trip to Jakarta. This customs agent had never missed a meal and immediately demanded that I pay him the equivalent of one hundred U.S. dollars for a special cat tax.

At this point in time, I was very tired and frustrated. However, I just smiled, reached out and shook his hand and said: "Congratulations, sir, you just bought yourself a cat!" I turned and walked toward my wife, who had a shocked look on her face.

The customs agent came running up to me, asking me to consider a lower cat tax. As I was angry, I sat down on the floor of the airport and negotiated with him for approximately thirty minutes. I got the cat tax down from one hundred U.S. dollars to less than two U.S. dollars. I don't think my wife or Snookiepuss appreciated my negotiating tactics.

In July 1985, President Suharto still had absolute control of the military and the government of Indonesia. The corruption and favoritism were rampant throughout the Indonesian government. When Barbara and I left Balikpapan, Indonesia, in 1982, President Suharto's wife, Siti Hartinah, was known as "Mrs. Ten Percent." In 1985, President Suharto's wife had a new name, "Mrs. Fifty-Fifty."

Barbara and I arrived in Jakarta the day the general manager for the Jakarta office retired from Conoco. The next evening the company held a retirement party for the general manager. At the party, I had the opportunity to meet the new finance manager. The finance manager enjoyed alcoholic libations and proceeded to give me the "scuttlebutt" on the retired general manager.

According to the finance manager, the former general manager was forced to retire due to several corporate improprieties. The issues included multiple conflicts of business interests. The former general manager was reported to own the office building Conoco leased in Jakarta. I was flabbergasted, as I couldn't imagine an expatriate manager with such unethical standards.

The finance manager also said the former general manager was notorious for inappropriate behavior with female employees. I thought the finance manager was just gossiping until I went to meet with the new general manager the next morning. The new general manager, a devout Christian, was standing outside his office watching with a shocked look on his face as workmen disassembled and removed a bed with "X-rated toys" from the back room of his predecessor's office. I, too, was shocked that an expatriate manger would exhibit this despicable behavior.

After a brief meeting with the new general manager, I met with the exploration manager. I thought the meeting would focus on the upcoming exploration program. However, I was told the Indonesian operation would be reorganizing and many nationals would be losing their jobs. Apparently, the company had increased the size of the organization in anticipation of exploration success in onshore Sumatra. The onshore Sumatra exploration program was an abysmal failure and Conoco was now forced to lay off employees.

The drafting department was reduced in size from fourteen to only one employee. I knew that most of the draftsmen depended on their jobs to provide basic necessities for their family. Fortunately, I received a phone call from the geologist that had been seconded to HUFFCO in Turkey. He was now exploration manager for his company's new operation in Indonesia and needed to hire an experienced draftsman. I was ecstatic to give him the names of all the draftsmen Conoco had just laid off. My old friend was very happy with the performance of the ex-Conoco draftsman. I made a friend for life with the ex-Conoco draftsman, who needed the job to put food on his family's table. I also gained the respect of many of our Indonesian employees for helping find the ex-Conoco draftsman a position with another company.

In Indonesia, expatriates were assigned national drivers. In Indonesia, especially Jakarta, the expatriate would always be at fault in any automobile accident. When Barbara and I arrived in Jakarta, I was assigned Robbi as our driver. Robbi had driven for the former general manager. I quickly found out that Robbi had picked up a few bad habits from the now retired general manager.

A few weeks after I started work in the Jakarta office, I was told that Robbi was very ill and was downstairs waiting for an ambulance. When I rushed downstairs, I saw Robbi lying on a stretcher waiting to be loaded into the ambulance. Robbi's eyes were closed, but I noticed he would occasionally open one eye and look around. After a few minutes, Robbi was taken by the ambulance to a local hospital, or so everyone thought.

There are no secrets in Indonesia. I quickly found out that Robbi's son had been to the office that morning looking for his father. Robbi had recently acquired a new girlfriend in addition to his wife, and his son was very angry. The son told all the Conoco drivers he was going to beat up his father to teach him a lesson. I also found out that Robbi told the ambulance drivers to take him to a private, Catholic hospital. Once the ambulance arrived at the hospital, Robbi told the Sisters he was Conoco Indonesia's general manager.

Robbi got into the private Catholic hospital, but the Sisters soon caught him in his lies. Robbi finally returned for work and apparently talked his son out of giving him a thrashing. However, I told Robbi that if he wanted to work for me, he had to stop the drama and focus only on driving. I told Robbi he would be terminated if he created any distractions for anyone at work. Robbi quickly became an exemplary driver.

Living overseas always provides the opportunity to learn new customs and cultures of the country. One of my more memorable cultural experiences was receiving an invitation for a circumcision celebration (Khitan). The Khitan was for the young son of one of my Indonesian colleagues. I learned that circumcision is considered a sign of belonging to the Islamic community. I was also told it was an honor for a foreigner, like me, to be invited to the Khitan. I had a wonderful time at the celebration and gained a little more insight into the wonderful Indonesian culture.

One of my first tasks as chief geophysicist was to ensure there were no financial improprieties in the department. Given Conoco Indonesia's history, I went through all the contracts with a fine-tooth comb. Only one contract for seismic data storage raised a red flag. The government owned all data that was acquired by any company in Indonesia. Companies were required to retain all geophysical and geological data until the oil and gas license was returned to the government. The company was required to turn over all the data in pristine condition. Failure to comply with the law could result in significant fines and even expulsion of the company from Indonesia.

Conoco had the majority of the seismic data stored at an established and secure facility. However, two seismic data contracts were with a company that had their facility in a residential neighborhood. The contract had been signed by "Mr. Happy," a national geophysicist, who now reported to me. The next morning, I told Robbi to take me to the data storage facility in question. Robbi immediately said: "Shouldn't you ask Mr. Happy to visit the facility? Inspection of a seismic data storage facility is beneath the chief geophysicist" Robbi's comment told me that I had an issue that had to be addressed.

The seismic data storage facility turned out to be an old, dilapidated home. After an hour's wait, the security person showed up to let me into the facility. The temperature in the home was close to 100 °F, and I could see mold growing on the boxes of Conoco's digital and paper data. I took a few pictures, much to the chagrin of the security person and Robbi.

When I returned to my office, Mr. Happy rushed up to me and breathlessly told me that he had shocking news. Mr. Happy told me one of the data storage companies wasn't fulfilling their contractual commitment. With my urging, Mr. Happy agreed to document the issues and terminate the contract with the data storage company. Mr. Happy also agreed to notify BKKA that all data storage issues would be resolved, immediately. My action sent a clear message to everyone in our Jakarta office.

Six months after I arrived in Jakarta, my boss told me I would be representing Conoco in a PERTAMINA sporting event. Remembering back to my experiences with the PERTAMINA basketball game in Balikpapan, I respectfully declined. My boss explained it was just a football (soccer) match between managers at PERTAMINA and managers with the international oil companies. I explained to him that I knew absolutely nothing about soccer. He said, "Exactly, that is why I picked you to represent Conoco, since PERTAMINA must win!"

I agreed to represent Conoco in the soccer game, since PERTAMINA was going to win—or, so I thought. Unfortunately, not all the oil companies had Conoco's cultural insight. A few of the European companies sent their fittest, most skilled soccer players to the match. The players from the oil companies were in their early to mid-thirties. The players from PERTAMINA were in their mid to late-fifties.

All the players met in the center of the soccer field, or pitch. The senior manager of

PERTAMINA welcomed everyone and concluded by saying. "Of course PERTAMINA must win." The European players laughed, which immediately drew a harsh stare from all the PERTAMINA players.

The fittest and most skilled European players started for the oil company team. Unskilled players, like myself, rode the bench. The European players quickly made several goals, repeatedly embarrassing the PERTAMINA players. At half time, the Oil Companies were leading 5 to 0, which clearly didn't meet PERTAMINA's pregame expectations.

In the second half, the unskilled players were given the opportunity to play. Shortly after I entered the game, one of the skilled European players stole the ball and broke down the sideline for what he expected to be a sure goal. I responded by sprinting down the opposite sideline. As I ran, I waved my arms at the referee, so he would see that I was offsides. I was flagged for being offsides, preventing another embarrassing goal being scored against PERTAMINA.

I was given an earful by the fit and skilled European players. A few minutes later, another skilled European player stole the ball and broke toward PERTAMINA's goal. Once again, I sprinted down the sideline waving my arms at the referee. Again, I was flagged for being offsides, saving another goal against PERTAMINA.

All the European players began shouting at me. Our team captain asked me, "Don't you know what you are doing wrong?" I responded, "Yes I do, but do you?" The European players stormed off, clearly frustrated with the American idiot.

The final score was PERTAMINA 0 and the Oil Companies 5. At the end of the game, I walked over to the PERTAMINA side of the field. I apologized to PERTAMINA senior manager for my team's poor form. He said: "Your performance was noted and appreciated." We chatted for a few more minutes and Robbi drove me home.

On Monday, my boss told me the senior manager from PERTAMINA had called about the soccer game. The PERTAMINA senior manager thanked my boss for sending someone to play in the soccer game that understood Indonesian culture. My friend, Toto, with BKKA told me that the final work permits for the fit and skilled soccer players who scored the five goals at the soccer game were not approved. I had learned a valuable lesson about Indonesian culture when I played in the infamous basketball game in Balikpapan. My former soccer teammates had also been provided a cultural lesson, but they were all on planes back to Europe.

Conoco's exploration history in Indonesia was abysmal, drilling over one hundred consecutive uneconomic exploration wells (dry holes). Global industry exploration success rates in 1985 ranged from 10% to 15% . Drilling over one hundred consecutive uneconomic exploration wells in Indonesia indicated there were major issues in all the technical evaluations.

My experiences with HUFFCO in Turkey taught me I couldn't fix a problem unless I thoroughly understood the root cause of the problem. As soon as I arrived in Jakarta, I began to review Conoco Indonesia's most recent license round evaluations in onshore Sumatra and offshore Java, as shown in *Figure 10*.

Figure 10

A technical risk assessment is a fundamental component in any license round evaluation. The technical risk factors in Conoco's evaluation were the presence of source rocks, reservoir, seal, and traps. The source rock evaluation estimates the probability of the presence of organic rich rocks in the region that can generate sufficient quantities of oil to fill subsurface traps. The reservoir evaluation estimates the probability of reservoir quality rocks that can produce oil at economic rates. The seal evaluation estimates the probability of rocks that can seal the oil in the traps. The trap evaluation estimates the probability of subsurface structures that have sufficient size to be economic, if filled with oil.

In my experience, the presence of source rocks is the most important technical risk component in a license round evaluation. If a region has no source rocks, every well in the region will be uneconomic. In a license round a company should only bid enough wells in a work program to prove or condemn the presence or absence of source rocks. Typically, two to three wells are sufficient to prove or condemn the presence of source rocks.

In each of the two licenses, Conoco's evaluation team accessed the presence of oil-prone source rocks to be almost a certainty. As a result, Conoco's technical team accessed the probability of finding multiple commercial oil fields to be almost a certainty. The technical evaluations resulted in Conoco committing to drill more than a dozen exploration wells in each license.

In the onshore Sumatra license, Conoco realized the lack of oil-prone source rocks after drilling only three or four exploration wells. In the offshore Java license, Conoco realized the lack of oil-prone source rocks after drilling only two or three exploration wells. In each case, Conoco was obligated to drill the remaining exploration wells in their license-round bid, wasting tens of millions of dollars. I then understood why Conoco's vice president of international exploration commented on the importance of providing a balanced presentation.

I thought the look-back analysis on the two failed license-round evaluations was important and could be used in Conoco's future exploration programs. However, my boss had no interest in looking at past failures. My boss wanted to discuss the enormous oil potential he saw in our existing exploration licenses.

In July 1985, Conoco Indonesia had one large offshore permit with one small producing oil field in the South China Sea and two onshore exploration permits in Irian Jaya, as shown on *Figure 11*. The terrain in the onshore exploration permit in the western region of Irian Jaya consisted primarily of swamp and marshlands. The terrain in the onshore exploration permit in the eastern region of Irian Jaya consisted primarily of mountainous, tropical rain forest. I knew that operating drilling rigs and seismic crews safely in Irian Jaya was going to be a challenge.

Figure 11

After the license-round look-back analysis, I focused on developing a low-risk, step-out exploration well near Conoco's lone producing oil field in the South China Sea. The well was a commercial success and gave the field a much-needed boost in oil production.

After the South China Sea exploration success, my focus turned to the exploration license in the western region of Irian Jaya. Conoco had discovered a small, uneconomic oil field, which was named the Wiriagar Field. Wiriagar was a shallow (3,000 Ft.) oil field in a carbonate reservoir.

Prior to my arrival in Indonesia, my boss had discussed exploration options with Conoco's senior management in Houston, Texas. One option was to drill thirteen shallow-exploration wells targeting the similar carbonate reservoirs as the Wiriagar Field. It would take two more oil fields, similar in size to the Wiriagar Field, to meet Conoco's economic threshold. The primary risk in this exploration option was whether oil-prone source rocks extended over the entire exploration license.

The second option was to drill two deep (10,000 ft. to 15,000 ft.) wells to evaluate the unexplored sedimentary section in the exploration license. One of the wells would test deeper objectives below the Wiriagar Field. The primary risk in this exploration option was whether the deep wells would find oil or gas. A natural gas discovery was only going to be economic if the reserves were sufficient to support the construction of a liquefied natural gas (LNG) plant, similar to HUFFCO's LNG plant in Bontang, Indonesia. Conoco's economic analysis indicated natural gas reserves had to be greater than four trillion feet of gas (TCFG) to build a new LNG plant.

This was not a simple exploration decision. Conoco Indonesia needed to make a significant new exploration discovery. Global oil prices were continuing to fall, which hurt every company's earnings. Oil companies began reducing exploration programs and closing unprofitable or marginal operations, like Conoco Indonesia.

My boss lobbied to drill the thirteen shallow-exploration wells. Conoco's senior management supported his recommendation, because drilling thirteen wells provided a higher probability of commercial success than drilling just two wells. However, the thirteen-shallow-exploration-well program assumed there were oil-prone source rocks over the majority of the entire license. It was possible there were no oil-prone source rocks outside of the immediate area of the Wiriagar Field. The first three wells in the thirteen-well-exploration program would prove the presence or absence of source rocks in the unexplored region of the permit. If the first three wells proved there were no source rocks, then it was almost a certainty that the remaining ten wells would also be dry holes. Unfortunately, my boss didn't want to discuss any option except economic success, which would prove to be an error in judgment.

In preparation for the thirteen-shallow-exploration program, the operations manager began to upgrade the operations base near the Wiriagar Field. In our weekly meeting, the operations manager complained that the national staff wanted him to sacrifice a white goat to celebrate the successful upgrade of the operations base. Our general manager said, "This is the accepted practice in Indonesia; what is the problem?" The operations manager replied, "I can't find a white goat!"

Two weeks later, the operations manager reported there had been an accident at the operations base. A national employee was smoking a clove cigarette in a no-smoking area. The employee threw the cigarette into a metal bin, which resulted in a small explosion. Fortunately, no one was seriously hurt. However, the operations manager would be spending the next few weeks writing reports on the unsafe actions of one of his staff. The operations manager would also have to develop additional safety procedures to reduce the risk of the incident happening again. This is the last thing he wanted to do, since he was focused on getting ready for the upcoming drilling program in the swamps of western Irian Jaya.

In our meeting, the operations manager explained that he had spray painted the goat white, because he couldn't find a white goat. He said all the nationals blamed him for the incident because he didn't locate a real white goat. As the operations manager paused, I said, "Make sure you put that the failure to find a white goat was the root cause for the accident." Everyone broke into laughter, except the operations manager.

The drilling department did an exceptional job of planning the thirteen-well program in the swamps and marshland of western Irian Jaya. The wells were drilled safely and efficiently. However, the first three wells were dry holes. More importantly, the first three wells had zero evidence of any source rocks. The results of the first three wells almost guaranteed that the next ten exploration wells would also be dry holes.

I went to my boss with a proposal to stop the shallow drilling program and take the remaining capital and drill one deep well in the Wiriagar Field. My boss's immediate reaction was shock that I would come up with such a crazy idea. He said the Indonesian government would never approve such a "silly idea." I countered that BKKA had just approved Mobil Oil doing a similar type of transaction in Kalimantan. Discussions with my friend Toto at BKKA confirmed that the Indonesian government wanted oil companies to find new oil and gas reserves. New discoveries meant more money for the government, while dry holes benefited no one.

My boss told me that he would not consider my proposal, and I should stop wasting my time on such "silly ideas." Unfortunately, the next ten exploration wells were all dry holes, as I had feared. The company relinquished the block in the western region of Irian Jaya. Another company acquired the license and drilled a deep exploration well in the Wiriagar Field. The exploration well was a major gas and condensate discovery, which is still producing.

Exploration's focus then shifted to drilling one exploration well in the eastern region of Irian Jaya. The terrain in the area was mountainous, tropical rain forest with an average rainfall of over two hundred inches per year. An exploration prospect had been developed with very limited geological or geophysical data. The probability of commercial success was very low, but the prospect had billion-barrel oil potential.

I flew out to the well location, prior to the drilling of the well. I flew by commercial jet from Jakarta to Sorong, Iran Jaya. I then flew by helicopter from Sorong to the well site. During the flight, the helicopter pilot asked if we wanted to see World War II relics. We landed on an airfield that had been abandoned by the Japanese at the end of World War II. The airfield had five Japanese fighters and bombers sitting on the grass runway, as if World War II had never ended.

After a brief tour of the abandoned Japanese airfield, we continued our helicopter flight to the well site. As we landed, I could see it was going to be a challenge to keep the rig resupplied because of the continuous rain and the low cloud cover. The region was also notorious for malaria and venomous snakes. Drilling this well safely was going to be a challenge.

We drilled the well and had encouraging results. However, we didn't have a definitive economic discovery. Conoco had to decide whether to drop the permit or to commit to drill another well to retain the exploration permit. I was asked to fly to Houston, Texas, to review the well results with senior management.

In 1986, the oil price was continuing to fall. Given the economic conditions, I decided to find the cheapest flight from Jakarta, Indonesia, to Houston, Texas. The Australian airline Qantas had just announced a new, incredibly cheap flight from Jakarta, Indonesia, to Los Angeles International Airport (LAX). Qantas would fly direct from Jakarta to Sydney, Australia. I was told I would spend twelve hours in a five-star hotel in Sydney and then fly direct to LAX. The round trip economy airfare approximately seven hundred U.S. dollars—approximately half of the next cheapest flight. Conoco's corporate travel policy allowed me to fly business class. However, I thought I would set an example for my staff to always look for ways to save the company money.

The Qantas flight took off on time, but within thirty minutes the pilot announced the plane would be landing in Denpasar (Bali). The flight attendant told me this didn't count as a stop as we were just picking up new passengers. After two hours on the ground, we finally took off for Sydney, Australia, or so I thought. Several hours later the pilot announced the plane would be landing in Brisbane, Australia. The amiable flight attendant told me, "No worries, mate, we'll get to Sydney on time." After another two hours on the ground, we took off for what I thought would be Sydney. One hour later, the pilot announced we would soon be landing in Adelaide. I was becoming concerned, as we were at least six hours behind schedule. The flight attendant also looked surprise about our landing in Adelaide. The flight attendant looked at me and said, "You got me, mate, but the pilot has everything under control."

We landed in Adelaide, spent another two hours on the ground, and finally took off for Sydney. We finally landed in Sydney more than eight hours behind schedule. I took a taxi to the Qantas five-star hotel, which turned out to be a seedy looking hotel in the seamier part of the city. I only had time to take a shower, change clothes, and catch another taxi back to the airport.

We took off from Sydney for what I thought would be LAX. Ten hours later, the pilot announced we would be landing at the Honolulu International Airport. We had to disembark while the plane was refueled. We disembarked the plane at 2 a.m. local time into a deserted airport terminal. After four hours, we were permitted to board the Qantas flight to LAX. After all the hours flying in economy class, my leg cramps had leg cramps.

As I settled into my seat, a very large Canadian gentleman took the seat in front of me. As the plane took off, the Canadian gentleman tried to force his seat back into a reclining position. In the process of forcing the seat back, he broke the seat and drove his seat into my knees. I let out a yell and the Canadian gentleman started cursing me.

The Canadian gentleman saw that I had fire in my eyes, so he said, "I'm sorry, these long flights from Hawaii to Canada make me irritable!" The flight attendants tried not to laugh. I just said, "Yes, I think I know all about long, tiring flights."

We finally landed in LAX over fourteen hours behind schedule. Fortunately, I was able to book a flight to Houston prior to the start of my meeting with Conoco senior management. A flight from Jakarta to Houston on most reputable airlines takes approximately twenty-four hours. My incredibly cheap Qantas flight had taken more than thirty-eight hours!

The meeting with Conoco's senior management went very well. Even with falling oil prices, senior management agreed to drill one more exploration well in the exploration license in the eastern region of Irian Jaya. After the meetings, I booked a flight back to Jakarta in economy class on another airline. I didn't think my body would take another thirty-eight-hour flight.

In 1986, Standard Oil Company of Indiana (AMOCO) acquired a large exploration license, adjacent to Conoco's exploration license, in the eastern region of Irian Jaya. A team of research geophysicists designed the parameters for AMOCO's seismic program in the mountainous rain forest of eastern Irian Jaya.

Oil prices were continuing to fall, but AMOCO spared no expense in the planning and design of the planned seismic program. My friend Toto, at BKKA, told me the estimated cost for AMOCO's seismic program was twenty-five million to thirty million U.S. dollars. I thought AMOCO could easily reduce their seismic program costs by ten million to fifteen million U.S. dollars.

I received a letter from AMOCO requesting permission to extend a few of their seismic lines into Conoco's exploration license. I immediately called my friend Toto since this type of request would require BKKA approval. I met with Toto and stated Conoco would agree to AMOCO's proposal if AMOCO would give Conoco all the data acquired in Conoco's exploration license. I explained this was standard operating procedure in other countries, like the U.S. and Canada.

I discussed AMOCO's proposal with Toto and his staff. I also provided Toto a model contract, which I had used in a similar situation in the United States. Toto told me that AMOCO's proposal was a new concept to BKKA. Toto said AMOCO's proposal was acceptable to BKKA because it would help assist exploration efforts in Indonesia.

I next met with my AMOCO counterpart and gave him my proposal. He quickly agreed with my proposal and we then discussed the details of AMOCO's seismic program. AMOCO had developed a theoretically sound seismic acquisition program. However, I quickly became concerned with AMOCO's geodetic process, which would be used to position AMOCO's seismic lines in relation to our two company's exploration licenses.

AMOCO planned to use a satellite global positioning system (GPS) to locate each seismic line. However, AMOCO hadn't taken the time to confirm the exact coordinates of their exploration permit and were using old paper maps provided by PERTAMINA. I pointed out there could be significant positioning errors using the old paper maps. My counterpart with AMOCO dismissed my concern and said, "No worries, it will be fine, Jack."

Several months later, my friend Toto called and asked me to come to BKKA, immediately. When I arrived, Toto told me AMOCO had shot almost 90% of the twenty-five-million to thirty-million-dollar seismic program in Conoco's exploration license. Toto couldn't believe that a large multinational company like AMOCO would make such a fundamental mistake. The chief geophysicist at AMOCO was promoted and transferred to Egypt. I have no idea if the real story of this multimillion-dollar mistake ever made it back to AMOCO's senior management.

After Irian Jaya, I focused on the exploration potential in Conoco's offshore license in the South China Sea. Conoco's research geochemists had just completed a source-rock study over the offshore exploration license. The research geochemists were virtually certain Conoco's offshore license had no remaining oil potential, but significant natural gas potential. These were the same people that did the ill-fated source-rock study for the onshore Sumatra and offshore Java Sea exploration license evaluation.

My boss developed a plan to explore, develop, and sell natural gas in Conoco's offshore South China Sea license to the Republic of Singapore. He put in many long nights to gain the necessary approvals from the governments of the Republic of Singapore, Republic of Indonesia, and Conoco's senior management. The gas program became my boss's sole focus.

In my evaluation of the offshore license, I found evidence for additional oil potential, not just natural gas. I explained my rationale to my boss and the importance of being prepared for different geological outcomes. The Republic of Singapore was only interested in buying natural gas for their power plants. An oil discovery could derail the gas sales agreement with the Republic of Singapore.

My boss dismissed my analysis, as I was a geophysicist and not a research geochemist. The first exploration well drilled in the South China Sea exploration gas program was a significant oil discovery. Once again, Conoco failed to appreciate the uncertainties and range of potential outcomes in a technical evaluation.

In June 1987, Conoco's corporate chief geophysicist visited our Jakarta office. I gave him a comprehensive update on all our geophysical programs. At lunch, he told me Conoco wanted me to transfer to the Netherlands. The company had just been awarded the "golden block" in the latest offshore licensing round in the Netherlands. The work program would include a massive 3-D offshore seismic survey. He told me they needed a technically strong chief geophysicist to go to the Netherlands. Barbara and I talked over the opportunity, and I accepted the transfer the next morning.

As Barbara and I were preparing to move from Jakarta, Indonesia, to Den Hague, Netherlands, the United States was producing 8.39 Million Barrels of Oil per Day (MMBOPD), while consuming 16.7 MMBOPD a day. The average price for oil in the United States in 1987 was $17.75 per barrel. The average price for gasoline in the United States in 1987 was $0.86 per gallon, which is equivalent to $1.61 per gallon, when the price is adjusted to inflation (March 2015).

From 1985 to 1987, the amount of oil produced in the United States decreased by more than 600,000 Barrels of Oil per Day (BOPD). The decrease in oil production was due to the expiration of the drilling tax credit policy and the fall in the global oil price. These two economic factors devastated the small American oil companies. Over the same time period, the consumption of oil in the United States increased by more than 900,000 BOPD. The increase in oil consumption was due to the fall in the price of fuel oil and gasoline in the United States.

Globally, the supply of oil was outpacing the demand for oil, which caused the dramatic drop in the oil price. Oil companies were now faced with volatile oil prices. The price volatility made revenue and earnings forecasts problematic. Most oil companies reacted by cutting overhead, eliminating exploration programs, and delaying nonessential projects.

In the United States, declining oil and natural gas prices undermined energy conservation programs and renewable energy projects. The world was awash with cheap hydrocarbons, and no one was concerned about energy conservation. The cost for electricity from wind and solar pilot projects was orders of magnitude more than coal and natural gas power plants.

The cost to build nuclear power plants in the United States was continuing to escalate due to increased safety regulations following the Three Mile Island partial meltdown in 1979. The 1986 Chernobyl nuclear disaster also crystalized opposition to further construction of nuclear power plants in the United States and Europe.

Global economies once again became accustomed to cheap oil and natural gas. This in turn would slow the development of renewable energy, like wind and solar. It would take more than a decade for the demand for oil to surpass supply. The next hydrocarbon shortfall would again result in the rapid rise in the oil price and renewed interest in renewable energy.

CHAPTER 10
Dutch Fool's Gold

In July 1987, Barbara, Snookiepuss (the cat), and I flew from Jakarta, Indonesia, to the Schiphol Airport outside of Amsterdam, Netherlands. Barbara and I didn't experience any of the Jakarta airport drama at the modern and efficient Schiphol Airport. It took Barbara only two weeks to find a beautiful, furnished apartment in Wassenaar, which is a suburb of Den Hague. My office was in Leidschendam, only a few miles away from Wassenaar.

Of all our overseas postings, the Netherlands was our favorite country. Barbara and I enjoyed the beautiful countryside, cycling paths, museums, and the engaging people. We also enjoyed the Dutch sense of humor. Shortly after arriving in the Netherlands, I asked a Dutch colleague where he would recommend my wife and I go for a good evening meal. His response: "France."

The headquarters for Royal Dutch Shell is in Den Hague, Netherlands. This company is closely associated with the country and the culture of the Netherlands. Like Elf Aquitaine in Gabon, Shell doesn't appreciate foreign interlopers on their "turf," especially since it is their home country.

In the Netherlands, the name of Royal Dutch Shell's operating company is Nederlandse Aardolie Maatschappij BV (NAM). NAM is a joint venture between Royal Dutch Shell (50%) and Exxon, now ExxonMobil (50%). I found this joint venture unusual, given Royal Dutch Shell's adversarial relationship with foreign companies in the Netherlands.

I was told the formation of NAM began with an exploration asset trade prior to World War II. Exxon had an exploration license over the entire country of Cuba and Royal Dutch Shell had an exploration license over the entire country of the Netherlands. The two companies agreed to an exploration asset trade. Royal Dutch Shell acquired a 50% interest in Exxon's exploration license in Cuba and Exxon acquired a 50% interest in Royal Dutch Shell's exploration license in the Netherlands.

No one could explain why the two companies agreed to the exploration asset trade. Perhaps each company identified significant exploration potential in the other company's exploration license. Whatever the reason, the exploration asset trade agreement was completed prior to the outbreak of World War II. As I was told, the existence of the agreement was either forgotten, misplaced, or destroyed as a result of World War II.

In 1959, Royal Dutch Shell discovered the Groningen gas field in the northeastern region of the Netherlands. The Groningen gas field is the largest gas field in Europe, with estimated ultimate recoverable reserves of approximately one hundred trillion cubic feet of gas (TCFG). After the Groningen gas discovery, a file clerk in Exxon came across the original agreement with Royal Dutch Shell. Exxon immediately notified what I suspect was a shocked Royal Dutch Shell. The final resolution of the original exploration asset trade agreement was the formation of NAM, a Royal Dutch Shell and Exxon joint venture.

I have never seen documentation that confirms this story. Is the story a fable or the truth that Royal Dutch Shell doesn't want to publicize? I do know that companies trading an interest in exploration licenses in different countries is an accepted practice in the oil and gas industry, even today.

The Dutch government has always encouraged the exploration for new gas reserves, even after the discovery of the Groningen Field. New discoveries are given a priority for gas sales over gas from the Groningen Field. The Groningen Field has become the swing gas producer in the country, making up for any shortfall in gas production from other fields. NAM produces over ninety percent of the country's daily hydrocarbons, even with the curtailment of gas from the Groningen Field.

Conoco opened an office in Leidschendam in 1983, after the discovery of the Kotter and Logger offshore oil fields. In 1987, Conoco was awarded two prospective offshore exploration licenses. Conoco was convinced these two prospective blocks contained multiple commercial gas fields.

Governments all over the world routinely hold oil and gas license rounds, which usually draw intense competition from many oil and gas companies. In most countries, companies submit sealed bid documents to the government, which include a work program for each license. The company with the most substantial work program is usually awarded the oil and gas license by the government.

The 1987 Dutch offshore licensing round included two unexplored licenses. The licenses had been a restricted military firing range. Producing oil and gas fields surrounded the two licenses, which the local press referred to as golden blocks.

The competition for the two golden blocks was intense. One block, Q-4, was considered to have the greatest hydrocarbon potential of the two blocks. On the Q-4 Block, Conoco, with two partners, bid five firm exploration wells and a 2-D seismic program. Another company, with three partners, bid four firm exploration wells and a one-hundred-square-mile 3-D seismic program. The Dutch government decided to award the Q-4 Block to the two partnerships that submitted the most significant work programs.

Conoco would be the operator of the Q-4 Block for the seven-company partnership. Large, multi-company partnerships are usually difficult as each company's strategy and budget can change from year to year. The work program for the Q-4 Block would include five firm exploration wells and a one-hundred-square-mile 3-D seismic survey. The estimated work program for the Q-4 Block was more than one hundred million U.S. dollars.

Conoco was also awarded the second golden block, Q-3. The work program on this license included two firm exploration wells and a thirty-square-mile 3-D seismic survey. Fortunately, we only had our two partners in the Q-3 Block.

Many companies have succumbed to the "winner's curse" phenomenon in a highly competitive exploration license round. The phenomenon is the tendency for the winning bid to exceed the intrinsic value of the item acquired. Winner's curse may occur for many reasons, including emotions, incomplete information, technical uncertainties, etc. Winner's curse means the winning bidder will face financial loss.

In 1987, the global exploration success rates for oil companies ranged from 10% to 15%. The low probability of commercial success confirms there is significant uncertainty in any exploration license. The probability of winner's curse occurring in any competitive global exploration licensing round is usually high.

Conoco experienced winner's curse in two Indonesian exploration license rounds. Conoco had bid very aggressively for the onshore Sumatra and offshore Java exploration licenses. Conoco's aggressive bids were due to limited data, unrealistic optimism, and the perception of intense industry competition.

As soon as I arrived in our Leidschendam office, I began reviewing the technical evaluations for the two golden blocks. Within a day, it was apparent the seismic interpretation had not been completed in a rigorous manner. The seismic interpretation had significantly overestimated the size of the structures or traps in the Q-4 and Q-3 Blocks. Overestimation of the size of the structures or traps inflated the potential reserves for each of the exploration prospects in each of the golden blocks.

Geologically, the reservoir assessment had not been completed in a rigorous manner, either. The reservoir properties were unrealistically optimistic. The over estimation of the reservoir properties coupled with the inflated size of the structures meant the reserve potential for each prospect was uneconomic! I notified my new boss, who thought I was being pessimistic. Once I went through my analysis with him, he agreed that it was possible the golden blocks were fool's gold.

The first step in the exploration work program was to acquire and process 3-D seismic surveys over the Q-4 and Q-3 Blocks. The final processed 3-D seismic data would provide a significantly better image of the subsurface than the existing 2-D seismic data.

The cost to acquire 3-D seismic data is significantly greater than 2-D seismic data. In the 1980s, 3-D seismic was usually acquired after a commercial oil or gas discovery, due to the high cost. However, the Q-4 and Q-3 work commitment stipulated the operator, Conoco, would first acquire the 3-D seismic data prior to drilling any well. I was pleased the 3-D seismic data would be acquired prior to drilling. I was confident the 3-D seismic data would resolve the differences between the initial Q-4 and Q-3 license round interpretation and my interpretation. I was hoping that the new 3-D seismic data would show that I was wrong.

The 3-D seismic survey in the Q-4 Block would be the largest 3-D survey ever acquired by Conoco. Our seismic acquisition specialists and the seismic contractor did an excellent job of planning, acquiring, and processing the two 3-D seismic surveys. Within one year, we acquired and processed two high-quality 3-D seismic surveys in the Q-4 and Q-3 Blocks on schedule and within budget.

In 1987, oil prices continued to fall, and oil price forecasts were also pessimistic. All companies were looking to reduce overhead and operating costs. Conoco mandated that all nonessential expatriates be repatriated to their home country. An expatriate usually costs two to three times more per year than a national in the home country.

Given the technical work that had been done on the golden blocks, I thought repatriating the expatriate geophysicists was prudent. However, the challenge was to find qualified Dutch geophysicists. The Dutch universities didn't produce many geophysics graduates, and the majority of the graduates went to work for Royal Dutch Shell. Employees at Royal Dutch Shell were reluctant to leave the company, as they knew they would always have a job. An American company didn't offer this type of job security.

Fortunately, I was able to recruit two highly qualified European geophysicists that wanted to live in the Netherlands. The geophysicists were from European Union (EU) countries, which meant they could work in the Netherlands. I was also able to recruit one Dutch geophysicist who was working for an American independent oil company.

The exploration manager, my boss, was transferred back to his home country for personal reasons. The new exploration manager was someone with whom I had worked when I first joined Conoco in London. He and I were both analytical and logical, which meant we could always work through any differences of opinions. I felt fortunate to have him as my new boss.

The advantage of 3-D seismic data is it provides a high-resolution image of the subsurface. The disadvantage of 3-D seismic data is that it creates an enormous volume of data that is virtually impossible to interpret without a sophisticated workstation. In 1988, two service companies, GECO and Landmark, dominated the global workstation market.

In our 1988 budget, I had included funds for three GECO workstations to interpret the new 3-D surveys on the Q-4 and Q-3 Blocks. We also acquired a large 3-D seismic survey over the Kotter and Logger oil fields. Our partners fully supported the purchase of the workstations.

My new boss told me Conoco's geophysical research and development group had developed a proprietary 3-D seismic workstation. I was asked to evaluate Conoco's proprietary workstation prior to making any workstation purchase.

I worked with three geophysicists and the information technology manager to develop a series of tests to compare the capabilities of the Conoco, GECO, and Landmark workstations. Within three weeks we had a unanimous verdict: purchase GECO workstations. Landmark had a quality workstation but didn't have GECO's technical support capabilities in the Netherlands. As to Conoco's preoperatory 3-D workstation, the verdict was not just no, but hell no! One geophysicist said he would quit if he had to work on the Conoco workstation. The other two geophysicists were more diplomatic but had similar opinions. The information technology manager's analysis was equally condemning of Conoco's workstation.

I wrote up a thirty-page report documenting the tests, results, and the analysis. I went through the report with my boss, who in turn wrote a supporting cover letter to the director of geophysical research and development and the region manager for European operations. The following week I was told the director of geophysical research and development would be flying to our office to review the report with me.

As soon as the director of geophysical research and development arrived, it was clear he wasn't interested in the analysis. His sole focus was to convince me to support the purchase of the workstations his department had developed. After an hour of going through the report in my office, he got up, closed my office door, and told me if I didn't support buying his department's workstation he would have me fired. My response: "I am paid to make technical and commercial assessments for Conoco, not make political decisions." With that response, he stormed out of my office and flew back to the United States.

Three weeks later, the region manager for European operations came to our office and asked to see the analysis on the workstations. After several hours, he said, "Why are we developing our own workstations?" My boss just smiled and said, "That is a question that should be asked by our senior management." Fortunately, I was allowed to purchase the GECO workstations and we commenced the interpretation of the new 3-D surveys on the Q-4 and Q-3 Blocks.

In 1989, we were ready to commence the Q-4 Block and Q-3 Block exploration drilling programs. Unfortunately, the 3-D seismic survey showed the size of the structures in each of the two blocks to be significantly smaller in size than Conoco's initial exploration license round evaluation. The size of the structures on the 3-D seismic were very similar in size to my interpretation, using the original 2-D seismic. There was no new data to change my assessment of quality of the potential reservoirs. My boss reviewed my work and agreed with my assessment: the golden blocks looked like fool's gold.

We held a technical meeting with our partners in the Q-4 and Q-3 Blocks. Our partners appreciated our candor and the rigorous assessment that we had done with the new 3-D seismic data. Although every company was disappointed, no one complained. Every company in the partnership had completed their own evaluation and every company had been overly optimistic in the assessment of the golden blocks, Q-4 and Q-3. To me, this proved that in a highly competitive exploration license round evaluation, it is very easy to be susceptible to winner's curse.

When we started drilling the first exploration well, I would go into the office to check the well results on the weekends and holidays. I wanted to know how the predrill prognosis compared to the actual well results and also look for any signs of oil or gas while the well was drilling. I knew Conoco's senior management and our partners wanted to be updated on any significant news as quickly as possible.

During my weekend work at office, I experienced one of many differences between the American and the Dutch cultures. The Dutch have a measured pace of life, while Americans tend to be in a constant hurry. This is especially true when it comes to dining.

I would usually get to the office by 6 a.m. on the weekends or holidays. After completing a quick check of the drilling results, I would go to a restaurant across from our office for breakfast. After breakfast, I would return to the office to do a more thorough review of the well results.

The first time I ordered breakfast at the restaurant, it took over two hours to get my order of two scrambled eggs, toast, and coffee. The next time I went to the restaurant, I spoke to the manager and explained my desire to get my breakfast quickly. The restaurant manager smiled and said, "We will try our best." This time it only took ninety minutes for me to receive my two scrambled eggs, toast, and coffee. Even though the restaurant staff tried to speed up my order, I could never get breakfast in less than one hour. I realized that I had to adapt to the culture of the Netherlands, as I was a guest in the country.

The drilling results on the first exploration commitment well in the Q-4 Block was, as expected, a dry hole. The first well had been drilled on the largest and best prospect. The probability of an economic discovery in any of the remaining four commitment wells was almost zero. The drilling results on the first exploration commitment wells in the Q-3 Block was also a dry hole. The golden blocks were indeed, fool's gold.

After the initial exploration well results, the partnership requested Conoco, as operator, to petition the Dutch government for relief from the remaining drilling commitment in the two golden blocks. Royal Dutch Shell had been an unsuccessful bidder on the two golden blocks. In a press statement after the license round, Royal Dutch Shell emphasized their commitment to the Netherlands. Royal Dutch Shell also emphasized the importance of fulfilling exploration license commitments, even when the results were less than expected. I knew that Royal Dutch Shell would make it very difficult for the government to release our partnership from the remaining drilling commitment in the Q-4 and Q-3 Blocks.

In 1989, Conoco completed a commercial agreement with several Japanese companies to fund a portfolio of global exploration wells. In the agreement, the Japanese companies would pay for 100% of the cost to drill over one hundred global exploration wells. The Japanese companies would earn 50% interest in any commercially successful exploration well. The total cost to the Japanese companies for the global drilling program was approximately 250 million U.S. dollars.

Several other American majors also negotiated similar commercial arrangements with other Asian companies. In principle, the arrangement met the goals of both parties. The American companies would gain capital for exploration wells, which would be deferred due to low oil prices. The Japanese companies would gain an opportunity to acquire new oil and gas reserves during a period of low oil prices. The Japanese companies recognized the delicate balance between global supply and demand and anticipated an increase in global oil prices in the next few years.

The technical staff and Conoco's European regional manager visited our office looking for exploration prospects to include in the portfolio for the Japanese companies. I showed them several low risks near field exploration prospects and higher-risk exploration prospects in the Q-4 and Q-3 Blocks.

After the meeting, I was told that all the prospects we presented were "too good" to be include in the portfolio with the Japanese companies. I asked about developing a balanced portfolio of high-risk and low-risk opportunities to ensure there would be some commercial success for the investing companies. I was told: "Don't worry, I'm sure we will have a few commercial discoveries."

The last well in the Japanese company investment program was drilled in 1991. Every single well in this drilling program was a commercial failure! The Japanese companies felt Conoco had not been forthright in the risk assessment of the exploration program. The reality was that Conoco didn't understand the fundamentals of exploration decision and risk analysis.

In 1990, I spent most of my time planning Conoco's first horizontal wells in the Kotter and Logger oil fields. I was also meeting with the Dutch government in an effort to reduce Conoco's remaining work commitment in the Q-4 and Q-3 Blocks. The meetings with the Dutch government officials were always cordial. However, I knew Royal Dutch Shell would make it very difficult for our partnership to achieve a reduction in the commitment work program for the two golden blocks. Conoco and the partnership had outbid Royal Dutch Shell for these two blocks in Royal Dutch Shell's backyard. I knew Conoco and the partnership were going to pay the penalty for trespassing in Royal Dutch Shell's home turf.

I always enjoyed the Dutch culture and their perspective on life. I was a member of a running club that organized a trip to France to run a ten-mile race in Paris. Although the race was great fun, I will always remember the run for the clash of the Dutch and French cultures. We had over eighty runners making the trip from Den Hague to Paris. Our hotel was approximately twenty miles from the starting point of the run. Our entire trip went as planned, until we all got up for breakfast at 5 a.m.

The hotel had promised to provide breakfast, which to the French meant one croissant and coffee. To the Dutch, breakfast meant at least two *pannekoeks* (Dutch pancake) with milk, juice, and coffee. An argument with the hotel staff ensued, resulting in the Dutch taking over the kitchen and making *pannekoeks* for everyone.

We all got to the starting line on time and had wonderful weather for the ten-mile run. However, I couldn't stop laughing during the run over the clash of the French and the Dutch cultures at breakfast. I always admired the Dutch for their good humor and their straightforward manner.

In November 1990, Barbara and I went to Morocco for a one-week holiday. During our stay, my boss called and said I was being offered a transfer to Lafayette, Louisiana. He recommended that I not accept the transfer. He said I would soon be offered a staff position in our headquarters office in Houston, Texas. A staff position meant I was being groomed for senior management.

I had spent three years working to extract Conoco and our partners from the exploration work commitments in the not so golden blocks. Although I was involved in the largest exploration 3-D survey and the first horizontal wells in my company's history, I honestly didn't feel I was adding value to the company. A corporate staff position might be good for my career, but I knew this wouldn't be the type of assignment I would enjoy. A transfer to the United States offered me the opportunity to gain knowledge in other disciplines, such as formation evaluation, deepwater drilling, and reservoir characterization. Although we enjoyed our time in the Netherlands, Barbara and I agreed to accept the transfer to Lafayette, Louisiana.

As Barbara and I were preparing to move from the Netherlands to Lafayette, Louisiana, the United States was producing 7.36 Million Barrels of Oil per Day (MMBOPD), while consuming 17.0 MMBOPD a day. From 1987 to 1990, the amount of oil produced in the United States decreased by over 990,000 Barrels of Oil per Day (BOPD). The rapid decline in the global oil price caused hundreds of small American oil companies to go out of business, which contributed to the dramatic decline in domestic oil production.

From 1987 to 1990, the consumption of oil in the United States increased by over 300,000 BOPD. The increase in oil consumption was due to the fall in the price of oil and gasoline in the United States. Once again, supply and demand would dictate the global oil price and the price would determine America's enthusiasm for energy conservation.

On August 2, 1990, Iraq invaded Kuwait. The average oil price rose from $17 per barrel in July 1990 to $36 per barrel in October 1990. The dramatic increase in the oil price was due to United Nation (UN) sanctions, which forbade any country from importing oil from Kuwait or Iraq. The UN sanctions reduced global oil supply by more than 4 MMBOPD.

The rapid rise in the global oil and natural gas prices prompted global utility companies to increase the use of inexpensive coal for power. The United States has significant coal reserves. Coal offered American utility companies a secure, sustainable energy source.

The global demand for inexpensive fuel prompted a surge in coal production in the United States. Coal production from subsurface mines was rapidly outpaced by open pit or strip mining. The shift from subsurface mining to open pit or strip mining was due to lower operating cost and the demand for lower quality coal. In 1990, coal production in the United States exceeded one billion tons, as shown in *Figure 12*.[16]

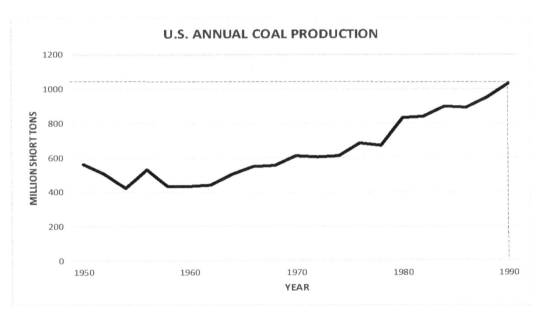

Figure 12

In the 1990s, the U.S. Congress began working on an energy policy. The legislation was designed to deregulate the utilities and create competition in the supply of electricity and natural gas to consumers. Competition drove the utility companies to reduce operating costs and to seek cheap, sustainable energy to fuel the power plants. The cost for nuclear and renewable energy, such as wind and solar, was significantly higher than coal, oil, or natural gas. As a result, utility companies continued to use coal as their primary fuel for power plants.

---

16   U.S. Energy Information Administration – Total Energy Review

CHAPTER 11

Roadkill Gumbo

Barbara, Snookiepuss the cat, and I arrived in Lafayette, Louisiana, in November 1990. We quickly realized that moving to Louisiana was comparable to moving to a foreign country. The gubernatorial election was the first sign of the state's unique qualities.

In 1990, there were three gubernatorial candidates: the incumbent Charles Roemer, former governor Edwin Edwards, and David Duke. The local newspapers characterized Charles Roemer as an honest person who wasn't personable, Edwin Edwards as a personable person no one trusted, and David Duke as a former Grand Wizard of the Ku Klux Klan.

My initial thought was the honest man, Charles Roemer, would easily win the election. It turned out to be a very close election with Edwin Edwards and David Duke drawing the most votes. In the runoff, Edwin Edwards won the 1991 Louisiana gubernatorial election against David Duke. In 1998, Edwin Edwards was convicted of racketeering, extortion, fraud, and conspiracy.

The next sign of Louisiana's uniqueness was my introduction to the Cajun culture. The "Cajuns" are descendants of French Canadians that settled in southern Louisiana in the late seventeenth century. You can still hear Cajun French spoken in parts of southern Louisiana today.

Shortly after I arrived in Lafayette, I went to meet each of Conoco Lafayette's supervisors and managers. When I went to meet the drafting supervisor, he was on the telephone. As I stood outside his office, I looked out a large window, which over-looked the company's parking lot. Just below the window was an old pickup truck with a huge turtle lying in the bed of the truck.

I said to no one in particular, "Look at that enormous turtle!" Fermin, one of the draftsmen, spoke up and said in a Cajun accent, "Mr. Jack, that is my turtle." I asked Fermin how he ended up with what appeared to be a dead turtle in the back of his pickup truck. Fermin told me picked up the turtle off the road when he was driving into work. He said he planned to make turtle gumbo.

I asked Fermin how long the turtle had been lying out on the road. Fermin told me with the winter weather there was no need to worry, as roadkill wouldn't spoil for at least four or five days. Fermin then asked if I would like some of his turtle gumbo. I declined, and told him I was allergic to turtle meat. I didn't lie, as I was sure I would be allergic to several-day-old, roadkill-turtle gumbo.

Barbara and I had moved so often we received Christmas cards from moving companies. The moving company that was handling our move from the Netherlands called me to tell me the sea shipment with our household goods would be docking in Houston, Texas. I was told his company didn't want to unload our household goods in New Orleans due to major theft issues at the port.

In less than four weeks, we bought a nice home in Lafayette; all we needed was our furniture. The moving company said our household goods had cleared customs and would be delivered to our home in two days. On the day the household goods were scheduled to arrive, Barbara told me she would handle the unpacking, and I should go to work.

Around 1 p.m. Barbara called me at the office and said she thought I should come home and talk to the movers. When I arrived home, it was a mess. Packing paper was everywhere and our antique furniture was strewn around the floor, waiting to be assembled. It was clear that this team of movers had no idea what they were doing. I asked the supervisor how long he had worked for his company. He told me over fourteen years, but his team made shipping crates and they had never unpacked a household shipment, before today.

I called the moving company's corporate office and told them of the problems. I told them I wouldn't authorize payment for the shipment unless they sent a qualified team to our home within the next twenty-four hours. Fortunately, the moving company complied, and we had our household goods unpacked, antique furniture assembled, and packing crates removed within the next forty-eight hours.

Although we had a lovely home and really enjoyed living in Lafayette, work was a very different situation. Conoco's U.S. oil and gas, or upstream, division operated very differently than the international upstream division. It was as if I was working for a different company.

Conoco had two different operating strategies for the U.S. and the international upstream divisions. In the U.S. division, the strategy was to drill over one hundred exploration wells a year. To implement this strategy the company employed over 1,200 geologists and geophysicists. The U.S. operating upstream strategy had not proven to be successful, which led to significant layoffs to reduce operating costs.

The international upstream division strategy focused on the value and risk of economic success. The international upstream division had achieved significant economic successes in the offshore regions of Dubai, United Kingdom, and Norway. The international upstream division employed approximately two hundred geologists and geophysicists.

The Lafayette office was responsible for exploration and production operations for onshore Louisiana and the western region of the offshore Gulf of Mexico, as shown in *Figure 13*.[17] The New Orleans office was responsible for the exploration and production in the eastern region of the offshore Gulf of Mexico. The Houston office was responsible for the exploration of the deepwater region in the U.S. Gulf of Mexico. At Conoco, water depths greater than five thousand feet were considered deepwater. Exploration operations in these water depths require special rigs (semisubmersible rigs or drill ships) and are usually very costly wells to drill. The Corpus Christi, Texas, office was responsible for exploration and production for the eastern region of onshore Texas. The Gulf of Mexico offices seldom shared technical information and were destructively competitive.

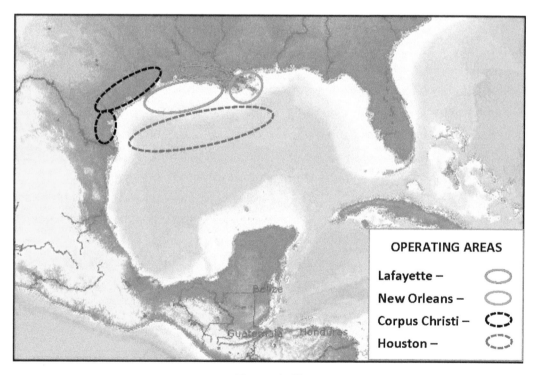

Figure 13[18]

---

17   Gulf of Mexico Foundation
18   PennEnergy PennWell Corporation Map

My new boss, the exploration manager, had an excellent reputation as a technically astute geologist, and was a kind and compassionate man. His compassion would make his job extremely difficult over the next few years.

Unfortunately, my fellow exploration supervisors were angry about my arrival. My peers saw me taking a job away from the U.S. upstream organization. One supervisor went out of his way to make my life at work as difficult as possible. This certainly wasn't the work environment that I expected when I accepted the transfer from the Netherlands.

On the positive side, my new exploration team had developed a three-well exploration program in the offshore Gulf of Mexico. The exploration wells were targeting deeper targets, below shallow producing oil and gas reservoirs. The prospects had been identified using high-quality 3-D seismic data, which was acquired primarily for the shallow-producing reservoirs. The advantages of 3-D seismic data were just being recognized by many companies in the Gulf of Mexico.

The spotlight was on me since this would be the first exploration drilling program by the Lafayette office in over eight years. The technical work on each of the three exploration prospects was thorough. I thought we would have at least one commercial discovery, based on my risk assessment.

The drilling results were better than I expected. We had two significant exploration discoveries, which created approximately two hundred seventy-five million dollars in net present value (NPV) for Conoco. Equally important, the two exploration discoveries would begin producing hydrocarbons within a few weeks. The new hydrocarbon production meant an important boost in revenue for the Lafayette office. The two discoveries more than replaced our office's annual production, which hadn't happened in over ten years.

Overall, the economics for the three-well exploration program was outstanding, or so I thought. I was asked to meet with the general manager to review the results of the exploration program one day when my boss was out of town. I thought the general manager wanted a high-level overview of the geology, economic results, and future production forecast. Instead, the general manager wanted to discuss the "failure," i.e., the one exploration dry hole, not the program's overall success.

To the general manager, drilling any well that wasn't a commercial success was unacceptable. I explained to him that commercial success rates for exploration wells drilled by other major operators in the Gulf of Mexico were less than 25%. We had just achieved a 67% exploration success rate, and more importantly, the economics showed the entire project, including the dry hole, were outstanding. Unfortunately, this was a concept that I was unable to explain to him.

After the meeting, I called a geologist I knew with the company in Houston, Texas. His first words to me were, "Congratulations on your two exploration discoveries. These two discoveries are the only good news our company has had in the entire Gulf of Mexico operation." I explained my perplexing meeting with my general manager. Fortunately, the geologist was a friend with my general manager. The geologist followed up with a call to my general manager to explain to him the reality of exploration risk and basic economics.

Three months after the exploration successes, my company's U.S. upstream division went through a major reorganization. The reorganization was designed to reduce overhead by laying off staff. Oil and gas prices were continuing to fall, and earnings were declining rapidly. This would be the first of many layoffs Conoco employees would endure over the next five years.

In 1993, Conoco moved all the Gulf of Mexico operations to Lafayette, Louisiana. This meant the closing of offices in New Orleans and Corpus Christi. The exploration team in Houston working the deepwater Gulf of Mexico would also move to Lafayette. The new office would be called the Gulf of Mexico Business Unit. The office consolidations took a heavy toll on the already low morale of Conoco's U.S. upstream division.

In 1993, the average price for oil in the United States was $16.75 per barrel. From 1990 to 1993, the price of oil had dropped to $6.44 per barrel. The United States was producing 6.85 Million Barrels of Oil per Day (MMBOPD), while consuming 17.7 MMBOPD a day. From 1990 to 1993, the amount of oil produced in the United States decreased by over 500,000 Barrels of Oil per Day (BOPD). From 1990 to 1993, the consumption of oil in the United States increased by over 300,000 BOPD. The decrease in oil production and the increase in oil consumption can be attributed to the decline in the global oil price.

The hydrocarbon industry was reeling from the dramatic drop in hydrocarbon prices. Conoco's reorganization meant another round of layoffs and organizational changes. I would no longer be a supervisor; instead, I would be the lone geoscientist on a commercial development team. The new position didn't give me a great deal of confidence that I had a future with Conoco.

The commercial development manager, my new boss, was a reservoir engineer. He was very familiar with the putrid financials of our Gulf of Mexico Business Unit. He also had a keen understanding of economics. I found it very easy to talk to him, as he was an analytical and logical person.

One of our first projects was to assess the economics of using natural gas, instead of gasoline, for our field vehicles. In 1993, natural gas prices were approximately 30% cheaper than oil prices. The cost to convert a truck in the field from gasoline to natural gas would pay out in less than one year. Engines fueled by natural gas run cooler than engines fueled by gasoline. As a result, engines fueled by natural gas have lower maintenance costs. The switch from gasoline to natural gas for our field vehicles was made as fast as possible. Every penny helped our Gulf of Mexico Business Unit's meager bottom line.

In 1993, the American Association of Professional Landmen (AAPL) launched the North American Prospect Exposition (NAPE) in Houston, Texas. The purpose of NAPE was to create a venue to bring prospect generators and investors together. I thought this event had the potential to dramatically shorten the cycle time required to develop a prospect, find investors, drill the exploration well, and produce first oil or gas. I discussed NAPE with my boss and he immediately saw potential for Conoco to sell some of their marginal fields to smaller companies.

NAPE was successful, exceeding almost everyone's expectations. Over three hundred million dollars in asset transactions were completed at the very first NAPE. Five years later, over one billion dollars in asset sales were completed at NAPE. The exposition's overwhelming success was due to several factors that were unique to the oil and gas industry in the United States.

As global oil prices fell in the 1980s, large, multinational oil companies began systematically laying off staff to reduce overhead. The never-ending rounds of layoffs meant even the large oil companies didn't have sufficient staff to properly oversee all their producing fields. Large companies began assigning their top performers to the most profitable fields and selling the less profitable fields.

A marginally profitable field to a large company was a gold mine to small company. Small companies didn't have the corporate overhead of large multinational companies like Royal Dutch Shell, Exxon, or Chevron. The small companies would assign their highly motivated staff on the newly acquired field and almost always turn a sizeable profit. NAPE provided the venue for sellers and buyers to meet and quickly close a deal.

Banks and investment firms also attended NAPE. Small, undercapitalized companies could now buy producing fields, using the proven oil reserves as collateral with the lenders. This allowed companies to continually sell producing fields to other companies. I know of one case where a field was sold seven different times before the field was ultimately depleted of oil. The final operators of this field were a husband

and wife who were engineers and lived only one mile from the small producing field in northern Louisiana.

The federal and state environmental agencies always kept a close eye on every producing asset sale. These agencies wanted to be certain the smaller companies maintained proper safety and environmental standards in the producing fields. The smaller companies almost always had very high environmental and safety standards. The field may have been an afterthought to the large company, but it was an exceptionally important asset to the small company.

At the first NAPE gathering, a few geologists were marketing low-risk exploration prospects they had developed using public domain well data. The geologists had purchased the mineral rights over the prospect for a few thousand dollars. The geologists wanted a company to drill their prospect in return for a small overriding royalty. After two days, all the geologists had successfully found buyers for their prospects. One geologist put up a sign that read, "All Prospects Sold, Gone Fishing."

Within a few years, NAPE became the most significant event in the U.S. oil and gas industry. Eventually, NAPE included international oil and gas opportunities. Foreign governments came to NAPE to advertise the oil and gas potential in their countries. Large multinational companies also began to sell international producing oil and gas fields at NAPE. In my opinion, NAPE helped to reinvigorate the oil and gas industry in the United States.

My boss didn't understand exploration, but he was willing to learn. He was especially interested in the concept of decision and risk analysis in exploration. A few weeks after I started explaining the concept of assessing exploration risk, my boss came into my office and said: "I think I may have made a mistake."

The Gulf of Mexico Business Unit consisted of old, producing oil and gas fields. Some of the fields had been producing for almost forty years. Many of the old fields were barely economic. As a result, the Business Unit's financials were rapidly eroding. Many of Conoco's other business units were also financially distressed.

Conoco had an oil field on the north slope of Alaska. This field had reserve potential of more 400 Million Barrels of Oil (MMBO). North Slope oil was transported in the Trans-Alaska Pipeline System (TAPS) to the port of Valdez in southern Alaska. Companies had to pay a high tariff to transport oil in the eight-hundred-mile-long pipeline system. The high pipeline tariff, operating costs for the field, and the low oil price made Conoco's field uneconomic to produce. Conoco was forced to shut-in the field and hope for the oil prices to increase.

British Petroleum (BP) wanted to acquire Conoco's oil field. BP was one of the largest oil producers on the North Slope and was also one of the owners of TAPS. BP's massive North Slope operations meant they could make Conoco's oil field profitable.

BP and Conoco discussed many potential asset trades. My boss negotiated a trade of the North Slope oil field for a 60 MMBO field in the Gulf of Mexico. The 60 MMBO field in the Gulf of Mexico was generating significant cash flow, which was desperately needed by Conoco's Gulf of Mexico Business Unit.

My boss told me he thought he had an agreement in place with BP. Near the end of the negotiations, BP demanded my company include all their deepwater exploration leases in the Gulf of Mexico. When my boss refused the proposal, BP walked out of the negotiations. My boss relented and agreed to BP's demand to include Conoco's numerous deepwater exploration leases. My boss asked me if I thought he had made a mistake.

The reality was Conoco needed the cash flow from BP's Gulf of Mexico oil field. It was also true that Conoco's deepwater exploration program had discovered only two marginally economic oil fields. The economics of Conoco's entire deepwater exploration program, including the value from the two oil fields, was a financial failure.

The exploration leases traded to BP were acquired over a new deepwater geological play. The reserve potential for the new play was significant. However, the cost to drill and develop even one field in this play would cost billions of dollars.

Did Conoco have the financial resources and commitment to successfully pursue deepwater exploration? I thought it was unlikely, given the exploration well costs and my company's balance sheet. I think my boss made the right decision given all the information he had available to him at the time. Unfortunately for Conoco, the traded leases contained the largest offshore oil discovery in the Gulf of Mexico.

In 1994, there was another reorganization in the Gulf of Mexico Business Unit. Along with staff reductions, we also had a new exploration manager. Our new exploration manager was charismatic, energetic, and very bright. My boss asked me to brief the new exploration manager on the current commercial development projects.

As requested, I gave the exploration manager a brief overview of the current commercial development projects. After my overview, the exploration manager asked me if I had read any good books. I gave him the names of the two books I had just read on business case studies. I told him my favorite book was *Don't Fire Them, Fire Them Up*. This developed into a lengthy discussion on exploration, business, and leadership.

After our discussion, the exploration manager asked if I could think of an action or symbol that would help light a spark in exploration. I suggested he put up a bell in the exploration conference room, which would be rung when there was a commercial exploration discovery. The team that developed the prospect would ring the bell and then make a presentation to the entire Gulf of Mexico Business Unit on the exploration success. I explained to him I was stealing the idea of the bell from my favorite book, *Don't Fire Them, Fire Them Up*.

I certainly enjoyed the meeting with the exploration manager. However, I didn't realize I was being interviewed for a new exploration supervisor position. I later found out that my boss on the commercial development team had recommended me for the new position.

A few weeks later, the exploration manager announced my appointment. This announcement wasn't warmly received, since I was still considered a member of the international group. Exploration moved to another floor and we needed to assign offices. We grouped people by their project teams and gave the best office to the most senior technical person. I intentionally took the smallest office, which had no windows. At a monthly exploration forum, someone complained about his office, and I responded, "You can take my office." That caused a laugh and helped create some sorely needed goodwill.

Performance appraisals were always a difficult topic, especially with the staggering number of layoffs. I will always remember doing an appraisal of a senior geologist who couldn't understand why he was receiving a "Met Expectations," when ten years earlier he had always received "Exceeded Expectations." I explained to the senior geologist he was at a much higher grade or position than he was ten years prior. A higher grade meant higher technical standards and level of performance. I also pointed out that in 1980, the U.S. upstream division employed approximately 1,200 geologists and geophysicists. In 1995, the U.S. and international divisions employed only seventy-seven geologists and geophysicists. If he hadn't been appraised an "Exceeded Expectations" ten years ago, he would not be with the company at the moment.

I believe my explanation helped the senior geologist understand his performance appraisal rating. I realized that I had become numb from the continuous reorganizations and layoffs. In fifteen years, Conoco had laid off more than 1,100 hundred geologists and geophysicists. It was then that I realized that Conoco's strategies were missing the mark.

In October 1995, I attended an executive management program at Penn State, entitled "Engineer/Scientist as Manager." I felt fortunate that Conoco sent me to this management program. The program had executives from six continents and a diversity of industries.

Dr. Ramani, professor in mineral engineering at Penn State, provided us with a most interesting anecdote of all the speakers. After Bill Clinton was elected president of the United States in January 1993, his wife, Hillary Clinton, started working on a government healthcare program. Mrs. Clinton organized a forum of renowned scientists and engineers to discuss how to provide government health care in the United States.

Dr. Ramani was invited to this meeting, as he was recognized as an innovator who had the ability to solve challenging problems. A reception was held the night before the start of the forum. Mrs. Clinton approached Dr. Ramani at the reception and asked him his thoughts on health care. Dr. Ramani said it is evident that cost was the single greatest issue with any government healthcare program.

Mrs. Clinton then asked if Dr. Ramani had any thoughts on how to reduce healthcare costs. Dr. Ramani responded: "I know how to reduce healthcare costs by fifty percent." Mrs. Clinton and her staff drew closer to hear Dr. Ramani say: "Helping people die two weeks sooner will reduce health care costs in the United States by fifty percent." Dr. Ramani was simply stating a fact that fifty percent of all healthcare costs are associated with prolonging a person's life by two weeks. Dr. Ramani concluded his story by saying Mrs. Clinton was on the opposite side of the room before he could finish his reply to her.

To me, the most insightful module in the management program was on adaptive and creative thinkers. We were given a test to assess how we approached a problem. The test scores ranged from one to one hundred and eighty. Adaptive thinkers had lower scores, while creative thinkers had higher scores. The test had absolutely nothing to do with an individual's intelligence.

After we took the test, the instructor divided up the class, grouping people with similar scores. We were then told to draw on one flip chart page (30 inches by 25 inches) our team's business model. I was paired with a Danish engineer. We had both scored between 125 and 130 on the test. We were given thirty minutes to develop and draw our business model on the flip chart.

A total of eight teams participated in the exercise. The team with the lowest scores, adaptive thinkers, gave the first presentation. Their business model drawing was a one-hundred-person organization chart with titles of each position perfectly printed in each perfectly drawn rectangle. The Danish engineer and I were shocked by the level of detail of what we thought was a conceptual problem.

The degree of creativity in the business model drawing slowly increased with each team. The Danish engineer and I gave the final presentation. Our flip chart contained a series of sketches showing the creation of an idea, a light bulb followed by decision points on whether to develop the new idea or sell the idea. The team that gave the first presentation had a look of absolute shock on their faces when they saw our flip chart. I am sure they thought we were quite mad.

The instructor then told us that people tend to hire within five points of their own score. He said, in a team setting, if one person has a score greater than ten points of the next closest person to him on the team, then that person will feel like the "odd person out" in any conversation. I thought this information was very helpful in optimizing team dynamics. To me, it also showed the importance of multiperson interview teams to get a broader perspective on any potential candidate.

On the final day of the program, we each had to summarize our thoughts and insights from the classes. In my opinion, the president of Consolidated Edison (Con Ed) gave the most memorable presentation. In 1995, Con Ed was one of the largest utilities in the United States.

The Con Ed president said the United States Energy Policy Act would soon come into effect. He said, prior to this management program, he thought Con Ed might have one or two new competitors in the northeastern region of the United States. The Con Ed president concluded by saying this management program helped him see that his company would have dozens of new competitors. He said his management team would have to develop a completely new strategy to address the changing dynamics of the utility business.

A few months after I returned from the executive management program at Penn State, the exploration manager resigned from the company. He had become frustrated with Conoco's bureaucracy and politics. He decided to start his own company. I was stunned by his leaving, but only wished him the best.

In 1995, there was another series of staff reductions along with the appointment of a new exploration manager. The new exploration manager was intelligent and hardworking, but her predecessor was a hard act to follow. However, I found it very easy to talk to her, since she was a logical person.

Conoco decided to return to deepwater exploration in the Gulf of Mexico. Conoco's deepwater lease position in the Gulf of Mexico was very limited following the BP asset trade. Conoco did have one exploration prospect in one of the small deepwater producing fields. The exploration prospect was significantly deeper than the producing zone in the small oil field. The exploration well was estimated to cost approximately one hundred million dollars. Conoco would have to pay 100% of the well cost, since the other partners were not interested in committing funds for this high-risk exploration prospect.

Although oil prices had increased slightly, the financials for our Business Unit and the corporation were still anemic. I was given the task of trying to find a new partner to share the financial risk in the expensive deepwater exploration well. The landmen and our scout provided me with an excellent analysis on competitor activity in the deepwater Gulf of Mexico.

The competitor analysis identified two companies that had previously abandoned deepwater exploration in the Gulf of Mexico and were now considering returning to deepwater exploration. Companies that implement a strategy to reenter an area they had previously exited are usually anxious to make an immediate and positive impact. Both companies had a significantly stronger balance sheet than Conoco.

One of the companies, Mobil Oil, immediately expressed interest in participating in Conoco's deepwater exploration well. I met with my counterpart at Mobil Oil to discuss the technical merits of the prospect, exploration risk assessment, and preliminary well cost. In a relatively short period of time, we completed an agreement, "subject to management approval." Mobil agreed to a two-for-one promote on the exploration well. A two-for-one promote means the other company will pay all costs to drill the exploration well up to an amount equivalent to the final authority for expenditure (AFE) for the exploration well.

I was certain I would have no problem getting Conoco's approval for this agreement, since this was an expensive, high-risk exploration well. Although this was a quality exploration prospect, there was no follow-up potential if the well was a discovery. Much to my chagrin, I was told that Conoco's senior management had rejected the proposed agreement with Mobil Oil.

Apparently, Conoco's senior management had fallen in love with the Gulf of Mexico deepwater exploration prospect. Conoco ranked all their global exploration prospects based on simplistic economic indicators such as risked net present value (NPV) and rate of return (ROR). The Gulf of Mexico exploration prospect had excellent indicators, but no follow-up or exploration play potential.

Early in my career, I learned that the value is always in an exploration play, not an individual prospect. An exploration play consists of multiple prospects with similar geological characteristics and risks. If the first exploration well is a discovery, the exploration risk on the follow-up prospects can be significantly reduced. In other words, a discovery in an exploration play is like the first business in a franchise. If the business works, you could have Starbucks, not just a successful coffee shop in Seattle.

I tried to argue my point, explaining why the value in exploration is almost always in a play, not an individual prospect. However, the decision was made, and I had to call Mobil Oil to tell them I couldn't gain Conoco's senior management approval. Four months later, the exploration well was drilled, and it was a dry hole. The well cost exceeded the initial AFE and had a very negative impact on my company's financial performance.

At the height of my frustration, I received a call from a recruiter at an executive search firm. He had an opportunity for a regional management position at another large multinational oil company. The company was Mobil Oil, the first company I worked for after university. Initially, I was skeptical as to whether Mobil Oil would be any different from Conoco. The recruiter told me Mobil Oil was very different from the time when I left in 1978 and recommended I go for an interview.

I agreed to proceed with the interview. The recruiter told me it would be a two-step interview process. The first interview would take place with five technical supervisors in Dallas, Texas. The interview would last approximately two hours. The entire team of technical supervisors would interview me at the same time.

Although the forum was the five technical supervisors on one side of the table and me on the other side of the table, I was up for the challenge. The questions from the interview team covered a broad spectrum of topics, including decision and risk analysis, technology, and leadership.

One of the questions centered on Conoco's decision not to farm-out the deepwater exploration in the Gulf of Mexico. I explained Conoco's decision was based on a prospect ranking system. As the prospect was drilled from an existing platform, the economics were robust. However, there was no follow-up potential in the license or the other leases in the immediate area.

I explained that exploration value is almost always in a geological play, not a single prospect. I gave examples of this concept in the North Sea, Indonesia, Gabon, and the Gulf of Mexico. I could see the technical supervisors understood the logic of my argument. However, it was also apparent this was a new concept to them.

The interview lasted over three hours. We had several good discussions, and I thought the interview went well. As I was leaving, one of the supervisors pulled me aside and asked, "How do you influence people?" He said this wasn't an interview question, but it was an issue with which he was dealing in his office.

I was a bit surprised by the question. However, I told him influencing people is almost always a challenge. Sometimes the issue can be related to different styles of communication or thought processes, such as creative versus adaptive thinkers. I concluded by saying influencing people requires you to be persistent and to continually look at different communication techniques.

Two days later, the recruiter called and said the first interview went very well. Mobil Oil now wanted me to meet the company's vice president of the U.S. business. I met the company's VP in a hotel lobby in Houston, Texas. The VP's first words were to thank me for my service in Vietnam. I thanked him and said that was only the fifth time I had heard those words since I returned from Vietnam in 1970. The VP had graduated from the U.S. Naval Academy in Annapolis, Maryland. He served five years in the Marines and then left the military to join Mobil Oil.

Our discussion quickly moved into topics on value creation, decision-making processes, risk, tactics, and strategies. The final question the VP asked me was, "Why do you want to leave your current company?" My response, "My analysis of a problem is always based on value. I believe my analytical approach isn't effective at Conoco. My goal is always to make a positive difference." Our discussion continued until we both had to catch flights home.

One week later, the recruiter called and said Mobil Oil wanted to offer me the position of Offshore Gulf of Mexico Region Manager, based in New Orleans. The final step was for me to meet with Mobil Oil's general manager for the Gulf of Mexico, who would be my new boss.

Two weeks later I met the general manager for the Gulf of Mexico region for dinner in New Orleans. The dinner discussion went well and concluded with him formerly offering me the position of region manager. I verbally accepted the position, as Barbara and I had discussed this potential move extensively. We agreed I would provide him a start date.

The next morning, I submitted my resignation to my boss in Lafayette, Louisiana. My boss asked me to reconsider my decision to leave Conoco. I thanked her but explained I had already accepted a new position and wouldn't go back on my verbal commitment. I expected my boss would want me to pack up and leave the office that day. Instead, my boss said she would like me to stay on and work for another three months! After a brief discussion, we agreed I would stay with Conoco for another thirty days.

Barbara and I have many fond memories of our time in Lafayette. However, one of the more humorous memories involves the wildlife that inhabits the town. Barbara and I would routinely see two or three raccoons walk along our wooden fence and jump down into our neighbor's vegetable garden, partaking of a garden buffet. We only had a flower garden, which was not at risk from the rampaging raccoons. Our neighbor had tried for months to catch the rapscallion raccoons. After numerous unsuccessful attempts, she finally gave up. The final score was Raccoons: 1 and Humans: 0!

One spring, Barbara noticed that an animal had been burrowing in our flower garden. We asked our neighbors, and they suspected the burrowing was caused by an armadillo. Barbara wanted this burrowing to stop, so she borrowed the trap our neighbor had unsuccessfully used for the raccoons. I went on the internet and found out that armadillos eat grubs and worms. Neither Barbara nor I were keen to dig for grubs and worms and late at night. Barbara decided that any self-respecting armadillo would certainly want to eat cat food.

That night, Barbara set the trap, expecting she would have an armadillo in the cage in the morning. Well, Barbara had an animal in the cage in the morning, but it was the neighbor's cat. This was very embarrassing for Barbara, since she is a cat lover. The final score was Armadillos: 1 and Humans: 0.

As my Barbara and I were preparing to move from Lafayette to New Orleans, the average price for oil in the United States was $20.46 per barrel. From 1993 to 1996, the price of oil had increased $3.71 per barrel. The United States was producing 6.46 Million Barrels of Oil per Day (MMBOPD), while consuming 18.3 MMBOPD a day. From 1993 to 1996, the amount of oil produced in the United States decreased by over 380,000 Barrels of Oil per Day (BOPD). From 1993 to 1996, the consumption of oil in the United States increased by over 1,000,000 BOPD!

The decrease in oil production was due to the industry's lack of confidence in the long-term stability of oil prices. The lack of confidence in long-term oil price stability resulted in the deferral of high-risk exploration programs. The dramatic staff reductions over the previous decade negatively impacted many companies' technical capabilities.

In 1996, the average price for gasoline in the United States was $1.23 per gallon, which is equivalent to $1.76 per gallon, when the price is adjusted to inflation (March 2015). From 1993 to 1996, the price of gasoline in the United States had increased by only twelve cents per gallon.

The federal and state governments in the United States attempted to legislate energy conservation. The automobile industry was mandated to improve gasoline mileage in new vehicles. New, improved, energy-efficiency regulations were also implemented on appliances and construction of new buildings. However, America was addicted to cheap, fossil fuel (coal, natural gas, and oil). Only dramatic increases in energy prices caused Americans to conserve energy.

CHAPTER 12
Big Oil in the Big Easy

In June 1996, Barbara and I moved from Lafayette to New Orleans, Louisiana. In less than four weeks, we found a nice townhome in the Garden District of New Orleans. We wanted to find a home close to work, since I expected to be spending many late nights at the office with my new company. The decision on our home proved to be a prudent decision.

New Orleans was a dramatic contrast to Lafayette, Louisiana. Lafayette had Red's, one of the top fitness clubs in the United States. Lafayette didn't have much of a nightlife, and virtually no crime. New Orleans, known as the Big Easy, was a city with a vibrant nightlife and a very high crime rate. Most New Orleans natives preferred eating beignets (pastry) to any form of fitness.

There were also dramatic differences between my previous company, Conoco, and my new company, Mobil Oil. Conoco's primary focus was cost containment. Mobil Oil's focus was on value creation, using decision and risk analysis. A focus on cost containment may provide short-term positive impact on the business unit's financial performance. However, overzealous application of cost containment can result in significant loss of value for the company. As an example, I saw another large oil company in the offshore Gulf of Mexico overlook a commercial discovery by failing to acquire the necessary data to evaluate the well. The company elected to not acquire data to save approximately seven hundred fifty thousand dollars on a well that cost more than fifteen million dollars. Another company acquired the lease, redrilled the same well, acquired the necessary data, and made a commercial discovery that had a net present value (NPV) of greater than two hundred and fifty million dollars.

Another significant difference between the two companies was the responsibility of the financial analysis. At Conoco, financial analysis on major projects was completed at corporate headquarters. At Mobil Oil, financial analysis of any project started with the team in the business unit. Mobil Oil's expectation was that everyone involved in a project understood the economic drivers. I thought Mobil Oil's approach to financial analysis was on the mark.

In my new position, I was responsible for the exploration and appraisal for the Outer Continental Shelf of the Gulf of Mexico, as shown on *Figure 14*. In my interview with the company's vice president of Mobil Oil's U.S. business, I said I thought there was a twenty percent probability I would recommend the company exit the region within the first year. I also said, I thought there was a seventy percent probability the company could achieve sustainable growth for a three-year period and a ten percent probability the company could sustain growth for more than three years due to new technology.

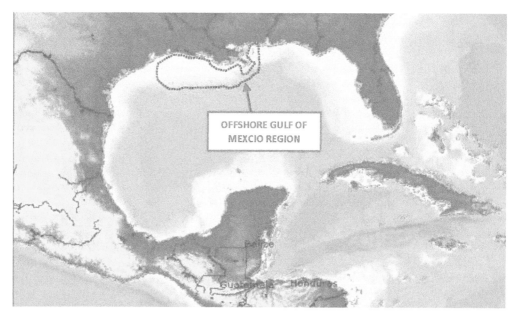

OFFSHORE GULF OF MEXCIO REGION

Figure 14[19]

When I arrived at the New Orleans office, my highest priorities were to 1) understand and rank the existing exploration plays, 2) evaluate and rank the prospect and lead prospect inventory, 3) meet and learn the capabilities of my team, 4) develop a comprehensive understanding of the budget and financial approval processes, and 5) understand the company's organizational framework and processes.

On my daily office walkabouts, I made a point of asking each person about a well

---

19   PennEnergy PennWell Corporation Map

that had just been drilled. My first observation was the team's penchant for documentation. I was however, surprised to find that few of the twenty-eight signatories on the well proposal knew much about the document they had signed. Even fewer knew about the results of the well that had just been drilled, including the well results, why the well failed, lessons learned, etc.

One of my first changes was to only have people responsible for developing the well proposal sign the document. The number of signatures on the well proposal went from twenty-eight to five. Everyone who signed the well proposal was expected to make a commitment to ensure every item on the document was correct and, if not, make the corrections immediately. Also, it was my expectation that the all signatories would closely follow the drilling of the well. I wanted to ensure that the drilling engineers had all the information necessary to make the best possible decisions while the well was being drilled.

I was impressed with the technical capabilities of the team. The team had identified six different exploration plays in our region. The next step was to rank exploration plays, based on the hydrocarbon potential, probability of exploration success, risked exploration economic analysis, and potential to grow our asset position in each play.

The team had never done a play ranking; therefore, I facilitated the ranking process. It was important that everyone had an opportunity to participate in the discussion. I wanted everyone's input, and more importantly, I wanted buy-in from everyone. The feedback on the process was very positive and I thought we were starting to work together as a team.

Over the weekend, I went to a local hardware store and bought a large metal bell. On Monday, I met with the team and explained the idea of the "discovery bell." I also explained the individuals who developed the exploration prospect would be asked to explain the results of the well and what the discovery meant to our region and our company. The overall feedback of these new processes was positive. I wanted to recognize the people that delivered economic success for the company.

Every three months, our vice president of the U.S. business held a quarterly review with his management team and all the region managers. Every region manager was required to provide an update on their business. Each region manager was asked to discuss their previous quarter, identifying any variances that impacted the economic performance of the business. Each region manager was also required to put forward a budget and work program for the next quarter and explain how the work program would impact the economic performance of the business.

In my first quarterly review, I summarized the top-rated exploration play and the budget and work program for the next quarter. One of our vice president's reports asked how I had selected this play to explore. I spent about twenty minutes going through the strengths and weaknesses of each play in our region and the ranking process. I concluded with the economic analysis to support my decision. When I returned to New Orleans, I was told I had passed the first test with Mobil Oil with flying colors.

We drilled the first exploration well and our initial analysis indicated we had discovered an uneconomic quantity of hydrocarbons. However, one of Mobil's technical specialists recommended we acquire additional data to ensure we didn't miss a discovery. He based his recommendation on an analog field approximately fifty miles south of our well location. As a team, we discussed the positives and negatives of our investment decision, and I approved the additional cost to acquire additional data.

The new data confirmed the exploration well had significantly more hydrocarbons than our initial analysis indicated. The appraisal or follow-up wells confirmed our region would soon have a new commercial field, producing hydrocarbons. Our region's goals were to add over 30 Million Barrels of Oil (MMBO) in new reserves and over 30,000 Barrels of Oil in daily production within three years. We met our three-year goals in less than eighteen months.

My next quarterly meeting went very smoothly. I left the meeting with full support for our appraisal program and our continued exploration program. However, I was already thinking about how to leverage this discovery into the next discovery. When I returned to New Orleans, I asked our technical team to look at all the dry holes within a forty-mile radius of our new discovery. I asked them to see if any company, focused on cost containment, could have missed a potential commercial discovery.

Two weeks later, the technical team returned with data from a well that had been drilled by a large oil company. The company had drilled the well, acquired a minimum amount of data, and had then abandoned the well. The public domain data showed this well looked just like the Mobil Oil discovery before additional data had been acquired. The large oil company and their partner still held the lease, which was approximately ten miles south of Mobil Oil's new discovery.

I set up a lunch meeting with the exploration manager with the company that had drilled the exploration "dry hole," approximately ten miles south of our new discovery. At lunch, I discussed my company's plan to focus on select plays in the Gulf of Mexico without giving any specifics. We discussed the diversity of prospective exploration plays that existed in the offshore Gulf of Mexico. We also agreed on the importance of focusing capital and personnel resources on the highest potential plays. I asked if his company would be interested in trading nonstrategic exploration leases in the Gulf of Mexico. The exploration manager was very receptive to my proposal and we agreed to have follow-up meetings.

In the follow-up meetings, we each provided the other with a map of our company's exploration leases in the Gulf of Mexico. We then exchanged a list of potential leases we would be willing to trade with the other company. On the other company's list was the exploration lease that was only ten miles south of our new discovery. The exploration manager with the other company then told me he wanted to acquire a specific lease from my company. I told him I would get back to him with a formal proposal.

Once I returned from this meeting, I asked the technical team to evaluate the exploration lease identified by the exploration manager with the other company. Within a few days, we were convinced we should proceed with the exploration license trade. I immediately followed up with a proposal to the exploration manager with the large company. He quickly agreed to my proposal, and I agreed to develop the exploration license trade agreement. Our legal and land departments worked several late nights to complete the exploration license trade agreement. Within another two weeks, we had successfully completed the exploration license trade.

Each company had only a fifty percent interest in the licenses that were traded. My next step was to get my partner in our new lease to either join us in an exploration well or to exit the leases. I set up a meeting with the exploration manager of the other company, who immediately told me his company was not prepared to drill another well on the exploration license. He added he was not prepared to exit the lease, because he thought my company had come up with a new exploration idea. He wasn't naive, as Mobil Oil had just acquired a fifty percent interest and operatorship in the exploration license.

I thanked the exploration manager for his time and returned to my office. Within two weeks, I had approval to prepare an Authority for Exploration (AFE) to drill an exploration well on our new exploration license. I met with the exploration manager with the other company and presented him with the AFE to drill an exploration well.

The exploration manager only had sixty days to respond to the AFE. If he approved the AFE, we would start drilling the well in the next ninety days. If he didn't approve the AFE, my company would pay for 100% of the exploration well. However, my company would also have 100% of a field if the exploration well was a commercial success.

The exploration manager told me again that he didn't have funds to drill this well. I reminded him that in the partnership agreement, his company had the right not to participate in the well. In this scenario, his company would not have to pay for any portion of the exploration well. However, if there was a discovery, then my company would have 100% of the field.

His only other option was to negotiate a farm-out, which would allow his company not to pay for the cost of the exploration well and retain a two percent interest in any exploration discovery. He agreed to my proposal and my company gained control of a very prospective lease.

The AFE committed my company to drill within ninety days or withdraw the AFE. The farm-out removed the one partner from the exploration drilling program and gave my company more time to select the optimum rig to drill the well. Our drilling manager thought he could secure the optimum rig to drill the exploration well in five to six months.

Our region had achieved significant economic success in a very short time. However, I was concerned about the remaining exploration potential in our region. After a thorough review with the team, I was convinced it was time to consider exiting the region. The company was ramping up the deepwater exploration program in the Gulf of Mexico. I was convinced the deepwater Gulf of Mexico region had significantly greater exploration potential and value than my region.

I met with the general manager, who concurred with my assessment. My general manager and I then met with our vice president to discuss our recommendation. Our vice president immediately said, "Good, because I have come to a similar conclusion." He had been in discussions with several companies on different commercial transactions. However, he thought one company offered the greatest potential to maximize the value from our company's assets.

A few weeks later, our vice president told me he had identified one company and the potential deal structure. In principal, our company would trade producing oil and gas fields and exploration licenses in my region and the onshore Louisiana region for a shareholder stock position in the other company. If the deal was completed, the vice president, commercial manager, and I would be on the board of directors of the other company.

In theory, this sounded like an excellent opportunity for my company to maximize value for nonstrategic assets. The people in our two regions would be redeployed to other strategic assets, such as the rapidly growing deepwater region. In practice, this was going to be a very challenging transaction to successfully complete.

The other company was a well-respected independent operator in offshore and onshore Louisiana. The independent operator would use stock in their company to buy my company's oil and gas assets. If information of this potential transaction were to become public knowledge, the independent's stock would skyrocket. If the independent's stock skyrocketed, Mobil Oil's purchasing value would be severely diminished.

My vice president told me I now had a new role in addition to my current position. I would lead the technical team in Mobil Oil's evaluation of the potential transaction. The stock price for the independent company was driven by the global oil price and their reserve assessment. I would be responsible for confirming that the independent company's reserve assessment was accurate. I would also be responsible for convincing the independent company that Mobil Oil's reserve assessment was accurate.

The independent company had been preparing their data room to support their reserve position for several months. I knew our reserve assessment for my offshore region would be able to stand up to a thorough review. However, the onshore Louisiana region didn't have a thorough reserve assessment in place. I was given only five days to prepare for the technical evaluation with the independent company. I knew I would not be able to meet the independent company's expectations for a comprehensive data room.

To further complicate the process, the independent company was based in Lafayette. Barbara and I had many good friends living in Lafayette. I was also known in Lafayette's tightknit oil and gas community. To further complicate matters, the independent company's office was located on the street opposite my former company, Conoco. It would be a challenge not to bump into people I knew from our days in Lafayette. Meeting with old friends and acquaintances would also increase the risk of rumors and drive-up the independent company's stock price.

It would be a challenge traveling unnoticed in and out of Lafayette. Mobil Oil set up a special data room for their onshore and offshore Louisiana oil and gas fields in Houston, Texas. I gave weekly status reports on our progress to the vice president and commercial manager.

The commercial manager and I flew to Mobil Oil's corporate office in Fairfax, Virginia to meet with the legal, financial, and investor relation team that would guide us in this potential transaction. I realized the complexity of the transaction when

I walked into a room filled with over forty legal, financial, and investor relation experts.

The independent company had a thorough and well-organized data room. I had an excellent team of technical specialists that I knew would be thorough in the assessment of the independent company's reserve assessment. I scheduled time every week to review in detail the status of the reserve assessment. I needed to know the potential weak points in the independent company's reserve assessment. It was imperative that I provide the vice president and commercial manager with a though understanding of the independent company's reserves.

My challenge was Mobil Oil's onshore data room and reserve assessment. With only five days' notice, our data room was not up to the standards of the independent company. I heard about the poor quality of Mobil Oil's data room at every meeting with the independent company. I suspected this was a ploy to try and reduce the amount of reserves Mobil Oil had in our producing fields. I was always cordial, since I knew if a transaction was going to occur, Mobil Oil's numbers would not be reduced, regardless of how many tears the independent company cried.

My week usually consisted of driving from New Orleans to Lafayette to meet with our team reviewing the independent company's reserve assessment. I stayed at a discount motel on the outskirts of town to minimize the risk of being seen by old friends or acquaintances. I then drove back to New Orleans to meet with my Gulf of Mexico offshore region team. I then flew from New Orleans to Houston to meet with our team reviewing our company's assets with the independent company and then flew to Dallas to provide the vice president and commercial manager an update on our progress.

The process went smoothly for eight weeks, when a local scouting service reported the independent company was a potential merger/acquisition target of a large multinational oil company. Mobil Oil was one of the companies listed as a possible buyer of the independent company. The independent company's stock went up a few dollars and then dropped back when nothing happened.

As 1998 was drawing to a close, there was one major merger. The merger would be with my company, Mobil Oil, and an even larger company, Exxon. Although it was called a merger, the reality was Exxon was acquiring Mobil Oil. The news sent shockwaves through Mobil Oil and the oil and gas industry. The merger would require approval from the U.S. government's Federal Trade Commission (FTC). However, industry analysts anticipated the FTC would approve the merger.

The pending merger resulted in a series of transfers throughout my company.

Mobil Oil wanted to place personnel in key positions prior to the merger. The U.S. deepwater Gulf of Mexico region exploration manager was transferred to Houston, Texas, to handle Mobil Oil's offshore Brazil exploration program. I would replace him as the region exploration manager in the deepwater. The transfer meant I didn't have to spend time on the potential transaction with the independent company. The transaction with the independent company was never completed due to the Exxon and Mobil Oil merger.

In my new role, I had an opportunity to work with some of the best technical professionals in my career. I will, however, remember one young geologist who had just joined the company from university. He came into my office with a training request to go to a four-week school on carbonate geology in the Bahamas. His role on our team had absolutely nothing to do with carbonate geology.

I looked at the training request and asked the young geologist to look out my office window. My office overlooked the company's parking lot. I asked him if he could see the old blue Ford Festiva in the parking lot? He said, "Of course I see that pile of junk." I said, "That pile of junk is my car and I spend Mobil Oil's money just like I spend my own money." He replied, "You could have just said no, just to save time." I smiled and thought, just saying no would not have been a teaching moment. Mobil Oil was going to be acquired by Exxon, but I wasn't going to waste company money.

In my new role as region exploration manager, I reviewed the deepwater Gulf of Mexico exploration plays and prospects with the team. Mobil Oil had spent over two hundred million dollars on deepwater exploration leases in the previous Gulf of Mexico lease sale. Much to my surprise, the prospect quality was very poor in the newly acquired leases.

I went to my new boss and told him my concern. My new boss had also been recently transferred into his position. His immediate reaction was that I must be a pessimist. However, he agreed to sit down with the team and review all the prospects. After only a few hours he pulled me aside and told me that he now thought I was an optimist. He concurred with my assessment that nothing in our deepwater exploration portfolio had the potential to be an economically viable prospect.

Our options were to recommend either Mobil Oil exit the deepwater Gulf of Mexico or find a new play that could provide material, economic reserves to our company. Our regional geological specialist identified two specific plays that could meet our company's economic criteria. Mobil Oil didn't have a significant lease position in either play. We would need to farm-in to another company's leases to pursue either play.

British Petroleum (BP) had a substantial acreage position in one of the priority plays, courtesy of my former company, Conoco. BP had recently acquired Standard Oil of Indiana (AMOCO) and was now capital constrained. Our landman confirmed that BP was interested in farming out a percentage of a well they would soon be drilling. I saw this as an opportunity for Mobil Oil to gain entry into an exploration play that had significant reserve potential.

Mobil Oil's senior management approved the proposed farm-in into the BP exploration well. In the discussion with senior management, I explained we could try and improve the terms of the farm-in, or we could try and acquire more offset or follow-up acreage in this high-potential play.

Mobil Oil's vice president of global exploration immediately stated he wanted to gain access to the play, not just one prospect. I had met with him prior to the meeting to provide him my assessment of our current deepwater Gulf of Mexico leases and the potential farm-in options with BP. He clearly understood the importance in investing in a play, not an inventory of isolated prospects.

With senior management's approval, we quickly entered negotiations with BP. In the meetings, it was clear that BP's Houston office didn't want a partner in this high potential exploration well. However, it was equally clear that BP's senior management in the United Kingdom had mandated the company would bring in a partner to reduce the capital exposure on a well that would cost over two hundred million dollars to drill.

As we were nearing the end of negotiations, BP suddenly changed the terms we had been discussing. BP demanded better terms than we had been discussing the previous week. When we declined the new demands, BP's team walked out of the meeting. My landman immediately said, "We must get approval to increase the terms or we will lose this opportunity". My response was, "First, we won't get approval to increase the terms, and second, I strongly suspect BP will come back in two weeks and agree to our initial proposal." I always suspected that my former boss at Conoco had been played by BP in the exchange of fields in Alaska and the Gulf of Mexico. In two weeks to the day, BP came back and accepted Mobil Oil's terms of the exploration well farm-in.

Although Mobil Oil and Exxon had agreed in principal to a merger, Mobil Oil could not discuss any aspect of this deepwater Gulf of Mexico exploration well with Exxon until the U.S. Federal Trade Commission (FTC) approved the merger. BP did an excellent job of drilling the well, which turned out to be the largest oil field in the offshore U.S. Gulf of Mexico. My friends at Conoco almost cried when they heard the news.

I was asked to meet with the FTC delegation to discuss my company's offshore exploration lease and producing asset position. The FTC's responsibility was to ensure that the merger would not create any unfair competition in any aspect of the oil and gas industry. The FTC team in New Orleans focused on onshore and offshore Gulf of Mexico for the two merging companies.

The FTC had no knowledge of other government regulatory agencies that oversaw the onshore and offshore oil and gas industry, including the federal lease sales. I showed the FTC a map of Mobil Oil and Exxon leases in the U.S. offshore Gulf of Mexico. Exxon's and Mobil Oil's combined lease position was less than five percent of the entire offshore leases in the U.S. Gulf of Mexico. The FTC was only concerned when two companies gained more than fifteen percent of the market share.

One of the FTC team asked if I could think of any area in the U.S. where one company had a dominate acreage position. I told him that the three companies BP, Exxon, and ARCO each had approximately one third of the Prudhoe Bay Field in Alaska, the largest oil field in the United States.

Two days later, BP announced plans to acquire ARCO. One day later, the entire FTC team left our office without any notice. Although the FTC approved BP's acquisition of ARCO, BP was required to sell ARCO's interest in the Prudhoe Bay Oil Field. ARCO's most valuable asset was the one-third interest in the Prudhoe Bay Field.

Neither Exxon nor Mobil Oil could notify their employees if they would have a position in the new company until the federal government approved the merger. When the federal government approval did come, it was a top-down notification process. First, senior management was notified of their new positions in the organization. The rumor mill was rampant with who was "in" or "out" in the new organization. I was told I would be formerly notified if I had a position in the new company within ten days.

Mobil Oil's vice president of global exploration came to my office, unannounced, and closed my office door. He said he couldn't formally notify me if I had a position in the new company, but he assured me I had made a very positive impression on Mobil Oil's senior management. He asked me if I still wanted to work overseas, and if so, would I consider returning to Indonesia. I told him Barbara and I enjoyed living overseas and we would enjoy returning to Indonesia.

He then said that Exxon had limited overseas operating experience, especially in developing countries. Mobil Oil had just made a major oil discovery in Indonesia. He said the new company, ExxonMobil, would need someone who was technically sound and culturally astute to go to Indonesia. He then pulled out a well log of the discovery well in Indonesia. He said, "I can't officially tell you if you have a position in the new organization, but you are only the fifth person in our organization to see this log." The well log showed a massive column of oil, and I knew it would be a very exciting project.

After the vice president of global exploration left my office, I began to think about the new company and working in Indonesia. Exxon would set the organizational framework and work processes for the new, merged company. My concern was that Exxon had a very different operating philosophy than Mobil Oil.

In the newly merged company, I would only be responsible for the technical elements of the exploration program in Indonesia. However, I knew the Indonesian government would expect me to work with them on both the technical and commercial elements of the project. I realized this would put me in a very difficult position with the newly merged company and the Indonesian government. I also realized I enjoyed integrating the technical and commercial components into any oil and gas project.

As I was pondering my future with the new company, I was contacted about a position with the Apache Corporation, an independent company. I met with the company's region vice president and received a job offer as the Gulf of Mexico exploration manager, based in Houston, Texas.

Professionally, I enjoyed my time with Mobil Oil. The company had dramatically changed from the Mobil Oil I joined in Denver, Colorado, in 1976. The technical professionals with whom I worked were some of the best I have ever known. I thought our vice president's business focus and style of management accountability were outstanding.

However, the newly merged company would be very different from Mobil Oil. I knew I would find it frustrating not to be able to integrate the commercial and the technical aspects of the business. If I couldn't fully support the new company's business processes, then it was time for me to leave the organization.

In 1999, as Barbara and I were preparing to move from New Orleans, Louisiana, to Houston, Texas, the average price for oil in the United States was $16.56 per barrel. From 1996 to 1999, the price of oil had decreased by $3.90 per barrel. The United States was producing 5.88 Million Barrels of Oil per Day (MMBOPD), while consuming 19.53 MMBOPD a day.

From 1996 to 1999, the amount of oil produced in the United States decreased by over 580,000 Barrels of Oil per Day (BOPD). From 1996 to 1999, the consumption of oil in the United States increased by over 1,200,000 BOPD! The decrease in oil production was due to the industry's lack of confidence in the long-term stability of oil prices. The lack of confidence in the increase in the oil price resulted in the deferral of high-risk exploration programs with the small, dynamic American oil companies.

In 1999, the average price for gasoline in the United States was $1.17 per gallon, which is equivalent to $1.60 per gallon, when the price is adjusted to inflation (March 2015). From 1996 to 1999, the price of gasoline in the United States had decreased twelve cents per gallon. From 1996 to 1999, the decline in the oil and gasoline prices had removed any incentive for Americans to conserve energy.

CHAPTER 13
Natural Gas Shortage Illusion

Barbara and I moved from New Orleans, Louisiana, to Houston, Texas, in October 1999. Houston was one of the fastest growing cities in the United States and suffered from horrendous traffic problems. Many people would spend three to four hours a day commuting to and from work. We wanted to find a home close to work, as I expected to be spending many late nights at the office with my new company. In a few weeks, we found a nice townhome near the Galleria area, which was just a few miles from my new office.

There were dramatic differences between my previous company, Mobil Oil, and my new company, Apache Corporation. Mobil Oil was a large multinational company with global upstream and downstream operations. Apache was an independent with global upstream operations that focused on acquiring producing assets. Although Apache had exploration programs in Australia and Egypt, the company's primary focus was asset acquisition in North America. Apache was recognized as one of the best in the world at extracting value from mature, producing oil and gas fields.

In the 1990s, most companies tried to identify and focus on their core skills and expertise. Majors like Royal Dutch Shell and British Petroleum acknowledged a lack of expertise at maximizing value from mature, producing fields. As a result, many major operators would actively sell packages of nonstrategic mature fields. Apache saw assets sales as a profitable business opportunity and developed a niche at maximizing the value from mature producing fields.

Another significant difference between the two companies was office demeanor and decorum. Mobil Oil was a coat-and-tie workplace that expected professional behavior from all their employees. Apache expected managers to wear a coat and tie, but the business conversations were always colorful—many times too colorful to my liking.

Organizationally, this was the leanest, flattest organization with which I had ever worked. My boss, the Gulf of Mexico region vice president, reported to the chief executive officer (CEO). The CEO reported to the chairman, who was still active in the day-to-day operations of the company. The chairman was one of the company's founders and was an innovative, outspoken entrepreneur.

Prior to my arrival, Apache had acquired twenty-two offshore producing oil and gas fields in the Gulf of Mexico from Royal Dutch Shell for more than seven hundred million dollars. This was one of many asset acquisitions that Apache Corporation completed over the next few years. Expertise in asset acquisitions allowed my company to develop a dominant oil and gas production operation in the shallower water depths (depths of less than six hundred feet) of the Gulf of Mexico.

Operationally, our region would drill from January through September and then release all the drilling rigs during the hurricane season in the Gulf of Mexico. The last three months of the year were used to develop new drilling projects for the next year. I arrived just in time to start work on the region's year 2000 work plan and budget.

At Apache, exploration meant any new well drilled in the immediate proximity of an existing oil or gas field. At Mobil Oil, these types of wells would be classified as appraisal or step-out wells. Appraisal or step-out wells have a very high probability of commercial success; however, the reserve targets are usually modest. My boss liked to say the key to our region's success was to "drill a heap and drill them cheap."

The exploration team consisted of only four geologists and four geophysicists. However, the team was very experienced, and their work ethic was off the chart. It was reassuring to me to see there was a robust inventory of quality exploration ready-to-drill prospects. I expected our region would drill approximately seventy-five to one hundred exploration wells in 2000. I was concerned about the intense nine-month drilling program with our limited staff. Each geologist and geophysicist team would be handling three to four drilling wells every day for nine months. The team would provide technical support to the drilling engineer to ensure each well was drilled safely and in the precise location. The team would also be involved in the identification of any hydrocarbons and commercial completion. Almost all of Apache's wells in the Gulf of Mexico region were commercial discoveries, which meant each team would be very busy. My boss told me not to worry, since this was Apache's way of doing business. He also explained that this was the reason the company had the best compensation package for technical professionals in the industry. In my new company, money was the primary motivator for management and the technical professionals.

My boss asked me to present the region's exploration budget to the CEO and chairman. In my presentation, I gave an overview of the geological plays we would be drilling and a summary of every prospect, including the geological risks and reserve potential. At the close of my presentation, the chairman stood up and said, "This is the best budget presentation I have ever seen." I really appreciated his comment, as it gave me, the new person in the organization, some much-needed creditability.

Apache had taken on significant debt to pay for the acquisition of the Royal Dutch Shell assets. Our 2000 region goals focused on rapidly increasing hydrocarbon production, eliminating nonessential operating cost, and exceptional safety performance. Smaller companies tend to be more safety conscious than many larger companies, as one incident will destroy their company.

At Apache, our individual bonuses were tied to the region's goals. My goals were associated with the addition of new reserves, annual hydrocarbon production, and safety. The production manager and I shared the same goals to ensure we worked closely together. The production manager and I became good friends, as we were both excited about the upcoming drilling and work program for the New Year. We also shared similar concerns on how we would handle the rapid increase in drilling activity. The production manager and I would routinely meet for lunch on the weekends to discuss issues from the previous week and future operational issues.

Our region goals were always tied to key financial metrics. Our primary goal was to increase production and pay down the debt. In 2000, Rate of Return (ROR) and Rate on Capital Employed (ROCE) were the most important financial metrics for our region. In January 2000, we began our work program with seven rigs. We had several initial successes, increasing reserves and, more importantly, production. The exploration team worked closely with the three-person drilling team. I quickly gained confidence the exploration teams would be able to handle additional rigs and maintain safe operations.

My first challenge was a former Royal Dutch Shell producing field in approximately seven hundred feet of water. One of the four exploration teams had identified several seismic responses, which could correspond to hydrocarbons. My concern was that Royal Dutch Shell's last well in this field was drilled on a similar seismic response and was a dry hole.

Unfortunately, my exploration team didn't have time do the type of quantitative geophysical analysis like I would have done at Mobil Oil. We either drilled one or more of these opportunities now or the rig would have to move to the next field. I discussed the risks and reserve potential with my boss and recommended we drill the best two prospects in the field.

These would be two of the most expensive wells we would drill in 2000, although the reserve and production impact could be significant for the company. The first well was a dry hole and I immediately started questioning my decision. The drilling manager, a crusty character, took great delight in announcing every dry hole at my boss's morning meeting. Fortunately, the second well was a commercial discovery. We continued to ramp up our drilling program. In June 2000, we had thirteen rigs operating in our offshore region, as shown on *Figure 15*.

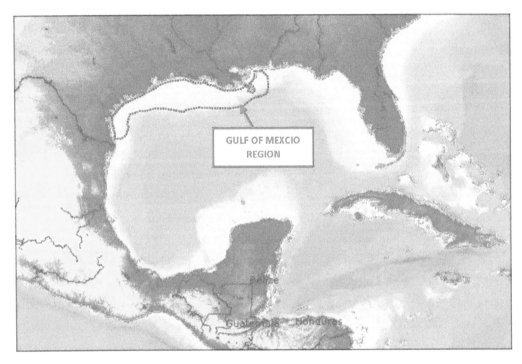

Figure 15[20]

At the end of the year, the region exceeded our production and reserve goals. The company responded with the largest bonuses I had ever seen in my career. I was shocked and very happy with my bonus. My boss asked me if I was motivated by money. My response was, "I am motivated by successfully completing a project, not the compensation." His response was, "You are weird." I smiled and thought my boss was probably right. However, his question caused me to question whether I wanted to stay with Apache for any length of time. Although the compensation was exceptional, Barbara and I enjoyed living overseas. I certainly enjoyed international

---

20   PennEnergy PennWell Corporation Map

exploration over the type of exploration I was doing at Apache in the offshore Gulf of Mexico.

One of the few disadvantages of working for this company was the "Apache Ten." This term referred to a ten-pound weight gain many people experienced after they joined the company. Apache was one of the most active companies in the Gulf of Mexico, and the service companies wanted our business. The service companies would routinely drop off enormous quantities of food in the morning and the afternoon to market their services.

In the morning, the coffee bar looked like a breakfast buffet with scrambled eggs, bacon, sausage, boudin, biscuits, pastry, fresh fruit, granola, yogurt, etc. The afternoon buffet consisted of sandwiches, wraps, tacos, pies, yogurt, fruit, etc. I needed all my discipline to keep from over indulging in the morning and afternoon buffets. I was also able to keep the weight off with my 4 a.m. runs at Memorial Park before I went to the office.

In 2000, I had the opportunity to participate in several asset evaluations and acquisitions. However, I will always remember one relatively modest asset acquisition. A company in Lafayette had a data room of all their oil and gas fields they were selling. I was asked to drive to Lafayette with a small team to evaluate several fields of interest.

Normally, it is a four-hour drive from Houston, Texas, to Lafayette, Louisiana. This trip took us over six hours to make the drive due to an accident on the interstate highway. While we were stuck in traffic, the reservoir engineer on our team decided to tell about the joys of watching professional wrestling. His stories were initially entertaining, but after six hours we all wanted him to shut up. Fortunately, the drive back to Houston, Texas, only took four hours and the geologist and geophysicist made a point of talking about the asset we had just evaluated, so we wouldn't hear any more about Hulk Hogan, Andre the Giant, and other wrestling celebrities.

The following week, I found a store that had life-size cardboard cutouts of different wrestling celebrities. I bought one cardboard cutout of Hulk Hogan and took it to the office. I had our draftsman put a picture of the reservoir engineer's face over Hulk Hogan's face. That night, after everyone had left the office, I put the cardboard cutout in the reservoir engineer's office. Everyone in our region was in on my prank. The next morning, as the reservoir engineer turned on his office lights, he was greeted by a life-size cardboard cutout of his smiling face on Hulk Hogan's body. The reservoir engineer was also greeted with applause from all his fellow coworkers. We all had a good laugh, which I think helped build a bit of goodwill in a stressful work environment.

The other reason I will always remember the asset evaluation in Lafayette was an absurdly inconsistent reserve audit on one field. Many companies use third party experts to do an independent audit on the reserves for their company. Accurate reserve estimates to an oil company are just as important as an accurate balance sheet is to a bank.

In the Lafayette data room, I discovered the other company used the same third-party reserve audit company as Apache. Apache was a partner with this company in one natural gas field. In theory, the third-party reserve audit company should have the same reserves in this field for both companies. However, the third-party reserve audit company had given 40% more reserves in this field to the other company than they had given Apache. This reserve discrepancy immediately set off alarm bells, and I called my boss in Houston, Texas.

The third-party reserve audit company had to explain to my boss and our CEO why there was such a discrepancy in reserves for the same field. The Society of Petroleum Engineers (SPE) has established rules and guidelines on oil and gas reserve assessment or certification. Although reserve audits are not an exact science, a forty percent variance was inexcusable.

In 2000, the price of natural gas started to rise rapidly, as shown in *Figure 16*.[21] Government experts assumed the price increases were associated with the demand for natural gas rapidly outpacing supply. In California, insufficient supplies of natural gas caused severe power shortages. Power shortages and high natural gas prices make voters angry and 2000 was an election year.

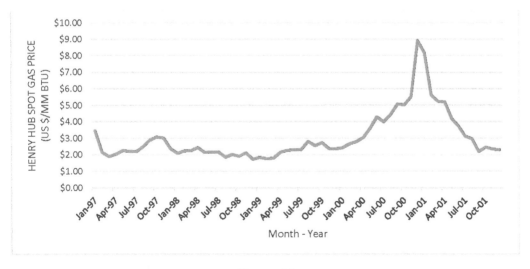

Figure 16

21   U.S. Energy Information Administration

The four largest natural gas producers in the Gulf of Mexico were asked to meet with the United States undersecretary of energy to discuss potential solutions to the natural gas supply shortfall. I was asked to represent Apache with the undersecretary of energy.

The four largest natural gas producers in the offshore Gulf of Mexico were Apache, two other independent companies, and a large oil and gas company. The two other independent companies proposed changes to U.S. tax laws that would improve the economics for their specific areas of interest. Their proposals were company focused, not country focused.

In my opening remarks, I stated the best way to increase natural gas production in any region is to drill more exploration wells. The offshore Gulf of Mexico was competing with other regions in the world for exploration dollars. The key question was what could the U.S. federal government do to make the offshore Gulf of Mexico more commercially attractive for exploration drilling? The three other companies ultimately agreed with me.

The undersecretary of energy asked if we could meet again in two weeks. He asked that we work together to develop different proposals that would increase the supply of natural gas for the United States. The undersecretary said he would bring senior bureaucrats from the Department of Energy to discuss our proposals. After the meeting, he pulled me aside and thanked me for helping to develop a consensus among the four companies to try and solve the natural gas supply shortage. I was cautiously optimistic about our next meeting with the undersecretary of energy. Given this was an election year, I expected that President Clinton would want to support changes to the tax laws that could address the natural gas supply shortage. Unfortunately, my optimism was misplaced.

The meeting started with a discussion on the global competition for exploration drilling dollars. I reviewed the number of offshore wells drilled in the world and then the number drilled in the U.S. Gulf of Mexico. The price of oil strongly influenced drilling activity. The number of exploration wells drilled always increased when there was an increase in the oil price. Falling oil prices always resulted in a decrease in drilling activity. However, drilling in the offshore Gulf of Mexico was steadily declining. The decline in drilling was due to the industry finding progressively fewer commercial drilling opportunities. Unlike other countries in the world, the U.S. federal government had made no changes to encourage companies to drill for the smaller exploration prospects.

After the initial discussion, I presented different ideas that could increase the drilling activity in the offshore Gulf of Mexico. The first idea was to introduce a sliding royalty to incentive companies to reduce the cycle time to drill exploration wells after the acquisition of the lease. The model was the faster the company drilled and brought a discovery into production, the lower the government take or royalty. The idea was to bring in any natural gas discovery as quickly as possible, which was one of the undersecretary's objectives.

This proposal brought immediate pushback from the chief economist from the Department of Energy (DOE). He pointed out that my proposal would reduce the government royalty. I countered that this incentive had the potential to make marginal discoveries economic. I presented analysis that showed the government's total revenue would increase with this proposal. It was apparent from the scowl on the chief economist's face that he was not going to support this proposal.

The next proposal was modeled after international licensing procedures used by most countries in the world. In the U.S., companies submit a sealed cash bid in an offshore licensing round and the highest cash bid wins the lease. In international license rounds, companies bid work programs to explore the block. The company with the most significant or impactful work program wins the license. I provided several papers by economic authorities stating that the international licensing round model was more effective for a country than the U.S. offshore cash bid system.

The chief economist again voiced even stronger opposition to this proposal. He said the U.S. government was forecasting the next offshore Gulf of Mexico lease sale would bring in over one billion dollars in revenue. I countered that increasing exploration drilling could deliver at least a four-fold increase in oil and gas revenue to the government. The revenue from the increased oil and gas production would exceed any licensing round revenues. The chief economist became very angry and stated the government's revenue from oil and gas production was confidential information. He then said I couldn't possibly make a reliable revenue forecast. I countered that annual oil and gas production volumes for the offshore Gulf of Mexico are published by the DOE. My calculation simply applied the standard government royalty rate on the government published production data. I said I suspected my revenue forecast was within plus or minus five percent of the confidential government numbers. The scowl on the chief economist's face grew even darker.

After the meeting, one of the members of the DOE told me that my revenue forecast was within one percent of the "confidential" government number. Unfortunately, it was very clear the chief economist and the senior government bureaucrats would oppose any change to the current offshore Gulf of Mexico royalty program. The undersecretary of energy thanked everyone for their participation. However, we heard nothing further from the undersecretary of energy and the DOE after this meeting.

Apache's chairman stated publicly that he saw no reason for the rapid price increase in natural gas. He thought the problem was due to either natural gas pipeline bottlenecks or trading irregularities. Many industry experts were skeptical of our chairman's opinion; however, he once again proved to be on the mark.

In 2002, *MarketWatch* reported that the June 2000 blackouts in California were directly caused by manipulative energy trading, according to a dozen former traders from Enron and its rivals.[22] The traders also stated that Enron Energy Services used the fear created by the blackouts to push large California businesses into more than one billion dollars in long-term energy contracts.

In 2005, *The Guardian* reported newly discovered tapes revealed how Enron shut down at least one power plant in California with the aim of raising natural gas prices.[23] The tapes revealed Enron's role in creating artificial power shortages in California, which caused residents billions of dollars in surcharges.

In the 2000 U.S. presidential election, George W. Bush defeated Al Gore. The Bush administration started the construction of Liquefied Natural Gas (LNG) import terminals on the coastal regions of the United States. The objective of the project was to import overseas LNG to supplement any natural gas supply shortfalls in the United States. There was no natural gas shortfall, and many of the LNG import terminals were later converted into LNG export terminals. The United States is now exporting LNG to Europe and Asia.

Apache made several more asset acquisitions in 2000. Our 2001 budget was even larger than our 2000 budget. I was very pleased with the prospect inventory that the exploration team had developed. I was confident our region would have another very good year. In January 2001, we began our work program with eight rigs. Our initial results were excellent, exceeding our early production and reserve addition forecasts. My boss was already looking forward to his end-of-year bonus check.

In May 2001, I received a phone call from the recruiting firm that placed me with Mobil Oil. The recruiter told me his firm had been contracted by an American independent company to find a senior exploration manager for their Malaysian operation. He asked if they could submit my name for consideration.

I asked for the company's name, company organization, and job description. As soon as I received this information, I researched the company, Murphy Oil, and reviewed their annual report. Barbara and I also did our research on Malaysia. Malaysia had been an English colony that had gained independence in 1957. In 2001, the prime minister of Malaysia was Dr. Mahathir bin Mohamad, who was a dominant political

22   Jason Leopold, *MarketWatch*, May 16, 2002
23   Julian Borger, *The Guardian*, February 4, 2005

figure in the political party, the United Malays National Organization (UMNO).

Under Prime Minister Mahathir's leadership, Malaysia experienced rapid modernization and economic growth. He was an advocate of third-world development and an international activist. However, the prime minister was an outspoken critic of many western economic policies. Unlike Indonesia, Malaysia didn't suffer from rampant corruption and favoritism.

Barbara and I discussed this opportunity and agreed I should follow-up with an interview. I met with Murphy Oil's vice president of international exploration. I thought my initial interview went well, although I sensed the interview process might take several months. I suspected the vice president didn't have the approval to extend a formal job offer for the position in Malaysia.

After the initial interview, I did additional research on Murphy Oil, which is headquartered in El Dorado, Arkansas. The company was listed on the New York Stock Exchange (NYSE) and consistently demonstrated strong financial performance. The CEO was a member of the Murphy family and had an excellent reputation in the oil and gas industry.

Over the next two months, I had three more interviews with Murphy Oil. I thought I had to be a finalist for the position, although the vice president liked to play his cards close to his chest. Another month passed before I received the job offer as Murphy Oil's senior exploration manager in Kuala Lumpur, Malaysia.

After I accepted the job offer, the vice president of international told me Murphy Oil and Apache had been at odds over a natural gas field in Canada. Apparently, the dispute resulted in the CEO's of the two companies having "words" with one another. The vice president recommended I not mention to Apache that I would be joining Murphy Oil.

The next morning, I submitted my resignation to my boss, who immediately wanted to know if I wanted more money. I told him compensation was not the issue. I simply wanted to do something very different from the work I was currently doing in the offshore Gulf of Mexico. This resulted in a meeting with the CEO, who asked me what he could do to change my mind. I thanked him and simply said it was time for me to leave Houston, Texas.

In 2001, as Barbara and I were preparing to move from Houston, Texas, to Kuala Lumpur, Malaysia, the average price for oil in the United States was $23 per barrel. From 1999 to 2001, the price of oil had increased by $6.44 per barrel. The growth of global economics, particularly in the People's Republic of China (PRC), was causing the demand for oil to increase. In 1993, the amount of oil produced in the PRC was equivalent to the amount of oil the country consumed. In 2001, the PRC imported approximately 1.6 Million Barrels of Oil per Day (MMBOPD). In 2016, the PRC imported approximately 7.6 MMBOPD.

In 2001, the United States was producing 5.8 MMBOPD, while consuming 19.65 MMBOPD a day. From 1999 to 2001, the amount of oil produced in the United States decreased by over 80,000 BOPD, while the consumption of oil in the United States increased by 130,000 BOPD.

In 2001, the average price for gasoline in the United States was $1.46 per gallon, which is equivalent to $1.91 per gallon, when the price is adjusted for inflation (March 2015). From 1999 to 2001, the price of gasoline in the United States had increased twenty-nine cents per gallon. The twenty-nine-cent increase was not an incentive for Americans to conserve energy. Car dealerships had waiting lists of buyers for gas-guzzling sport utility vehicles (SUV).

In 2001, the coal industry in the United States was still experiencing production growth, as shown in *Figure 17*.[24] The steady increase in coal production was due to increases in U.S. consumption and exports to Europe and Asia.

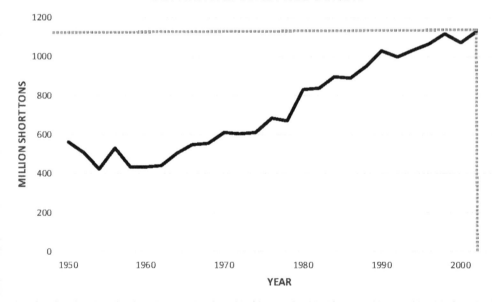

Figure 17

24   U.S. Energy Information Administration – Total Energy Review

As U.S. coal production was increasing, the quantity of high-quality coal was decreasing, as shown in *Figure 18*.[25] The highest to lowest quality coal is anthracite, bituminous, subbituminous, and lignite. The higher the coal quality, the higher the heat content from the coal.

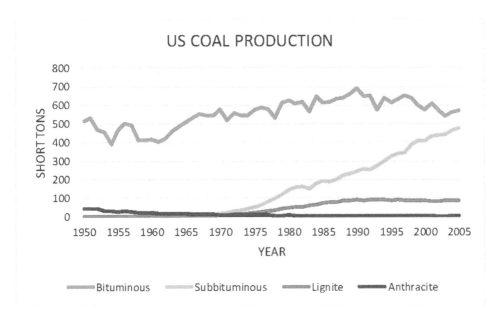

Figure 18

In 2001, China's coal production began a dramatic production increase, as shown in *Figure 19*. Approximately ninety percent of China's coal was used in domestic power plants for electricity. Today, China is the largest producer and consumer of coal in the world.[26]

25 Coal mining in the United States, Wikipedia
26 Coal China, Wikipedia

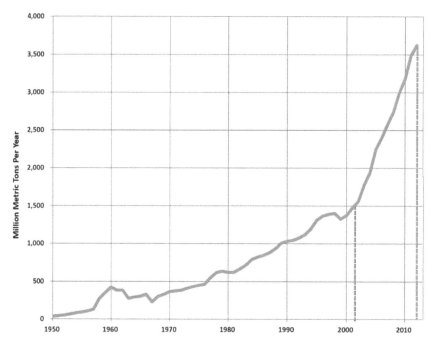

Figure 19

In November 1981, Barbara and I spent three weeks in China on holiday. On our first morning in Beijing, I went for a run. There were no cars on the streets, just bicycles and pedestrians. Initially, I thought I was running on a very foggy day. I soon realized I was running in thick, toxic smog. The smog was produced from home furnaces that burned low-grade coal.

In 1981, Beijing's population was reported to be 9.2 million people. In 2015, Beijing's population was reported to be 21.5 million people. In 1981, there were virtually no cars on the roads in Beijing. In 2015, Beijing reported having over 5.5 million private vehicles.[27] Today, the air pollution in Beijing can only be described as toxic.[28]

In 2001, global demand for all fossil fuels had grown significantly due to low price and abundant supply. Coal was the cheapest of all the fossil fuels and experienced the greatest production growth. In 2001, developed and undeveloped countries alike sought abundant supplies of cheap energy.

In the United States, nuclear energy saw the closure of older plants and slow growth in the construction of new power plants. In 2001, electricity from nuclear power plants had virtually plateaued, as shown in *Figure 20*.[29]

---

27   www.livefrombeijing.com/2009/11 Beijing's-vehicle-population
28   www.borepanda.com/pollution-china
29   U.S. Energy Information Administration

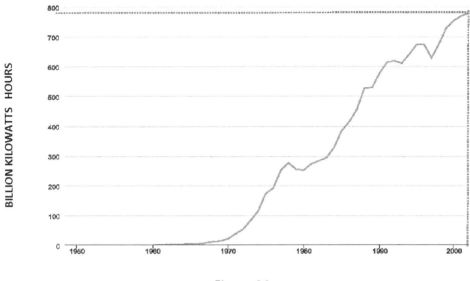

Figure 20

In 2001, the cost for electricity from nuclear power plants was significantly greater than the cost for fossil fuel. Nuclear power plants also experienced significant downtime for refueling and maintenance. In 2001, future growth in nuclear power in the U.S. was tenuous at best.

In 2001, the utility industry was skeptical of the renewable energy potential for electricity generation in the United States. The utilities skepticism was due to concerns over cost in $/kWh and long-term reliability. Most types of renewable energy were susceptible to the whims of Mother Nature. Energy storage in batteries was very limited.

The utilities skepticism of renewable energy would begin to change over the next decade. The utilities paradigm change was caused by the rapid price increase in all fossil fuels and major developments in wind and solar technology. Pollution from low-quality coal, especially in the PRC, would also motivate the utilities to pursue renewable energy.

CHAPTER 14

Kikeh, A Magical Fish

I started work for Murphy Oil in their Houston, Texas office, while my work permit was being processed. It took PETRONAS, the Malaysian national oil company, only four weeks to approve my work permit. I spent my time in the Houston office learning about Murphy Oil, the corporation, and the operating conditions in the country of Malaysia, as shown in *Figure 21*.

Figure 21

Murphy Oil had upstream operations in North America, Great Britain, Ecuador, and Malaysia, and downstream operations in North America and Great Britain. In many of the upstream operations, Murphy Oil was a nonoperating partner. As a non-operator, Murphy Oil could monitor the operator's technical work with a minimum of technical staff, minimizing overhead.

The exploration program in Malaysia signaled a shift in Murphy Oil's upstream strategy. The company operated two offshore exploration leases in the "shallow" waters (one hundred feet to two hundred feet) of Sarawak and two exploration leases in the deepwater (three thousand feet to nine thousand feet) of Sabah, as shown in *Figure 22*. As operator, Murphy Oil could control the pace and rate of expenditures of the work program. The operator also had the advantage of charging the partnership for all technical work done on the licenses.

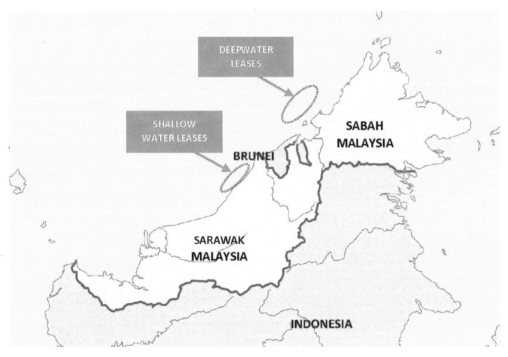

Figure 22

Another shift in Murphy Oil's upstream strategy was the equity or working interest in the exploration licenses. In North America, Murphy Oil's working interest in exploration leases ranged from 20% to 35%. A lower working interest allowed the company to reduce capital exposure on high-risk exploration projects. In Malaysia, Murphy Oil's equity was 80% to 85%. Murphy Oil clearly saw Malaysia as an opportunity that could transform the company. Time would prove that Murphy Oil made the right decision.

In Sarawak, Murphy Oil operated two blocks, which had previously been leased by Royal Dutch Shell. Royal Dutch Shell had released the two blocks after drilling numerous, unsuccessful exploration wells or dry holes. Murphy Oil saw exploration potential that had been overlooked by Royal Dutch Shell.

Murphy Oil had drilled an oil discovery on the first exploration well in the Sarawak shallow water block; however, the second well was a dry hole. In 2001, Murphy Oil's work program was designed to confirm the commerciality of the initial oil discovery and to drill one additional exploration well in one of the two Sarawak leases.

In Sabah, Murphy Oil operated two deepwater blocks. One block was awarded to Murphy Oil by PETRONAS and the other block was acquired from ExxonMobil. In 2001, Murphy Oil's work program was to acquire a massive 3-D seismic survey in each of the two deepwater blocks. The plan was to drill the first deepwater exploration well in March 2002. The acquisition of the 3-D seismic programs had already started prior to my arrival in Malaysia. The 3-D seismic acquisition was forecast to be completed by August 2001. As I reviewed the work programs, I was concerned about the limited time my team would have to develop a ready-to-drill prospect inventory before drilling began in deepwater Sabah, Malaysia, in March 2002.

Prior to my departure, the vice president of international told me I would not report to him, but to the general manager (GM) for Malaysia. The GM was an attorney who was an experienced negotiator. The vice president of international then told me I had a very difficult job, because he thought he was one of the world's best exploration managers. My first thoughts were "I have gone from the frying pan into the fire," and "I must be chopped liver."

In July 2001, Barbara, Ziggy (the new cat), and I flew from Houston, Texas, to Kuala Lumpur, Malaysia. Barbara and I didn't experience any of the Jakarta, Indonesia, drama at the modern Kuala Lumpur International Airport. It took Barbara only one week to find a beautiful, furnished apartment in Kuala Lumpur. Our apartment was near the magnificent PETRONAS Twin Towers and was within walking distance of my new office.

Barbara and I had visited Kuala Lumpur on holiday in 1981. At the time, Barbara and I found Kuala Lumpur to be a quiet, laid-back city. Since then, Malaysia experienced rapid modernization and economic growth, under Prime Minister Mahathir's leadership.

In 2001, Murphy Oil's Kuala Lumpur office was on one floor of a large multistory skyscraper. When I arrived, our office had approximately thirty employees. Our GM had worked for Murphy Oil for over twenty-five years and had an in-depth understanding of the company's history. Our GM's direct reports were the operations manager, finance manager, and me. The operations manager was a knowledgeable reservoir engineer who had over thirty years of experience. The finance manager was an experienced accountant who had never worked in an overseas operation. The GM and all his managers were American citizens, which would soon be relevant.

When I arrived in Kuala Lumpur, my exploration organization consisted of an expatriate exploration manager and four national geologists and geophysicists. The exploration manager was soon repatriated to our Houston, Texas, office. The national staff were all educated in the United States and had gained excellent training at their previous employer, ExxonMobil. I was confident our small team could handle the Sarawak exploration drilling program.

Petroliam Nasional Berhad (PETRONAS) is considered the custodian for Malaysia's national oil and gas resources. PETRONAS is a fully integrated, multinational oil and gas company. In 2001, PETRONAS was one of the top five hundred revenue-generating companies in the world. [30]

In Malaysia, PETRONAS's upstream organization consisted of PETRONAS, the administrative organization, and PETRONAS Carigali, the oil and gas company. PETRONAS's role was to ensure that all companies complied with the terms of their oil and gas licenses and conducted safe operations. In Malaysia, PETRONAS Carigali was usually a nonoperating partner that was assigned to new oil and gas ventures.

In Sarawak, Murphy Oil had only one partner, PETRONAS Carigali, with a 15% nonoperating interest in the two exploration licenses. After I arrived in Kuala Lumpur, I met with Murphy Oil's exploration team to assess the potential of the two exploration licenses. In a few weeks, we developed a drilling program to appraise the initial oil discovery in the West Patricia Field and drill one exploration well. The exploration well was designed to evaluate an uneconomic oil discovery drilled by the previous operator, Royal Dutch Shell.

In August 2001, my company contracted a jack-up rig for our Sarawak drilling program. Unfortunately, the drilling manager had not joined the company, which stretched our limited resources. The first well in the West Patricia Field encountered mechanical problems, and the well had to be abandoned prior to reaching the reservoir objective. The jack-up rig then moved to the exploration well location.

The exploration well was structurally up-dip or higher than the Royal Dutch Shell well, which encountered oil. Murphy's well encountered multiple reservoirs with natural gas, which had no economic value at that time. The pressure of having two

---

30   Fortune Global 500 - 2016

unsuccessful wells was felt by everyone in our organization and the vice president of international.

The jack-up then moved back to the West Patricia Field and drilled a successful oil well. Additional drilling would confirm a commercial oil field. I know everyone in our organization let out a huge sigh of relief with the confirmation of our first commercial field.

Royal Dutch Shell discovered oil in onshore Sarawak, Malaysia, in 1910. ExxonMobil discovered oil in offshore Peninsular Malaysia in 1969.[31] In 2001, Royal Dutch Shell and ExxonMobil dominated the private sector's upstream and downstream operations in Malaysia. These two companies strongly influenced PETRONAS's perception of the capabilities of the private sector. In 2001, Royal Dutch Shell and ExxonMobil were both pessimistic about the remaining exploration oil potential in Malaysia. PETRONAS would soon learn that Royal Dutch Shell and ExxonMobil weren't infallible.

Our GM and the operations manager met with PETRONAS following the successful drilling program of the West Patricia Field. Our operations manager stated that he thought Murphy Oil could complete the construction of the West Patricia Field, producing oil facilities, within three years. The senior manager for PETRONAS laughed and said, "Royal Dutch Shell couldn't complete the construction of these types of facilities within five years. How can your small company complete this development in three years?" Murphy Oil completed the construction of the West Patricia Field in only eighteen months. More importantly, the construction and operation of the West Patricia Field exceeded PETRONAS's safety, and environmental standards.

September 11, 2001, is a day that will live in infamy for the United States of America. Kuala Lumpur, Malaysia, is twelve hours ahead of New York City, New York. Barbara flew out of Kuala Lumpur, Malaysia International Airport, at 2 p.m. local time on September 11, 2001 to visit her parents in Edmonton, Alberta, Canada. That evening I turned on the television to watch the English language news services when the first plane crashed into the World Trade Center. A few hours later, Barbara's parents called to ask if I knew if her plane had landed in Vancouver, Canada. The United States had grounded all nonmilitary flights and information was almost nonexistent. Calls to the United States and Canadian embassies provided me with no information. These were very unnerving days for every American at home or overseas.

Barbara was supposed to fly from Kuala Lumpur to Hong Kong, change planes, and then fly to Vancouver, Canada. The local travel agent emphatically told me that

---

31   A Barrel Full Oil & Gas, Wiki

Barbara's plane had landed safely in Vancouver, Canada. However, Barbara's parents were told her plane had not landed in Vancouver, Canada. Another forty-eight hours passed before I learned Barbara's plane had been diverted to Japan. The passengers were housed in very basic accommodations without any access to a telephone or computer. Barbara finally arrived safely in Edmonton, Alberta, Canada, six days after she started her trip.

In September 2001, the acquisition of the deepwater seismic program in Sabah, Malaysia, was completed. However, the processing of the 3-D seismic data was almost two months behind schedule due to logistics delays associated with increased airport security due to the terror attacks in the United States.

Murphy Oil signed a contract to drill four wells in one of the deepwater Sabah, Malaysia, licenses. The semisubmersible drilling rig was forecast to arrive in Malaysia within the first two weeks of March 2002. The delays in the seismic processing meant our exploration team would have less than seventy days to develop four ready-to-drill prospects. My company's aggressive drilling schedule was making our job almost impossible.

Fortunately, I was able to supplement the deepwater exploration team with two experienced expatriate geophysicists and two experienced national geophysicists. However, only one expatriate and one national geophysicist had any deepwater exploration experience. Every geophysicist kept asking me how we were going to successfully complete this project in less than seventy days. Although I kept saying I knew we were up to the challenge, I was very concerned.

Our vice president of international was continuing to turn up the pressure on me to ensure we would be ready to drill in March 2002. It was important that I not pass the pressure down to the team or else our team would implode. Once the 3-D seismic data was loaded on our workstations, I asked the four geophysicists to spend the first two day "running their fingers through the data." Running their fingers through the data meant I wanted them to scan the large volume of 3-D seismic data and identify any interesting or obvious structures. I would then sit with the team and select the best lead or idea to be mapped out in detail by one of the four geophysicists. My approach seemed to remove some of the pressure from the deepwater team.

After two days, I sat down with the deepwater team to look at all the possible ideas or leads. We spent the next two days reviewing all the leads the team had identified. After the review, we had one obvious lead, which was then given to one geophysicist to develop into a ready-to-drill prospect. The remaining three geophysicists repeated the process for the next area in the massive volume of data.

My exploration strategy was to drill four prospects, which would evaluate different geological plays. If we drilled four straight dry holes, then I thought we would have effectively condemned this exploration license. If we drilled a discovery, then we could quickly shift to drilling appraisal wells and then drill other prospects in the play that we had just proven to be commercial.

Our vice president of international was convinced we would have multiple deepwater commercial oil discoveries. Apparently, Murphy Oil's CEO shared our vice president's confidence. However, no wells had ever been drilled in our very large deepwater license. Realistically, there was a low probability of finding a commercial oil field in our deepwater exploration licenses. The odds were almost insurmountable that I would be able to meet my senior management's lofty expectations.

The drilling manager arrived in Kuala Lumpur just in time to prepare for the Sabah deepwater drilling program. Fortunately, he had extensive deepwater drilling experience in the Gulf of Mexico. Our exploration team was starting to develop an inventory of ready-to-drill prospects. However, I still hadn't located an operations geologist who would interface with exploration, drilling, and the service companies. When the rig is drilling, the operations geologist is working twelve to twenty-hour days to ensure critical data is effectively communicated to all key personnel.

After looking at dozens of resumes for an operations geologist, I learned that an old friend with whom I had worked in Balikpapan, Indonesia, had retired in Kuala Lumpur, Malaysia. I was able to convince him to come out of retirement to help us drill this exciting, deepwater exploration program. He had decades of geological operations experience in Southeast Asia and worked very well with drilling engineers and service companies.

An experienced petrophysicist joined our exploration team from our company's London office. Although he had no deepwater experience, he would prove his worth in the development of the West Patricia Field in Sarawak. I was also able to recruit an experienced national geophysicist from Royal Dutch Shell to lead our offshore Sarawak exploration program. He was a highly competent professional, which allowed me to spend more time with the deepwater exploration team.

In January 2002, our limited resources were stretched with the ongoing development of the West Patricia Field in offshore Sarawak and preparation for the offshore deepwater exploration drilling program in Sabah. The five-day work week became a six- and then a seven-day work week. The long hours began to wear on everyone. I could see the pressure in the face of our normally personable GM. The drilling manager was known to arrive at his office early in the morning, close his office door, and scream, "I hate living in Malaysia!"

The deepwater exploration team had developed only two of the required four ready-to-drill exploration prospects. The drilling time for each prospect was approximately twenty-five days. I was concerned we were running out of time to develop the remaining two prospects within our limited window of time.

In February 2002, we met with our partner, PETRONAS Carigali, to discuss our deepwater exploration program in Sabah. Our exploration contract with the government required my company to pay for or "carry" our partner's cost on the first exploration well. After the first well was drilled, our partner, PETRONAS Carigali, would pay their twenty percent working interest on all future work. The partner meeting went very well, as PETRONAS Carigali was excited about the deepwater exploration potential in Malaysia.

I found out from my friends at PETRONAS Carigali that Royal Dutch Shell was convinced that "absolutely zero economic quantities of oil or gas exist in deepwater Sabah, Malaysia." I spent the next two hours discussing global deepwater exploration and explaining why our permit had excellent oil and gas potential. I couldn't help thinking that if Murphy Oil had a discovery, Royal Dutch Shell's mystique would be severely tarnished.

In March 2002, the semisubmersible rig arrived on schedule and we commenced drilling the first exploration prospect in the northeast region of the Sabah exploration license. The drilling of the well went very well, until the marine-riser system failed. Without the riser, the drilling mud would spill out onto the seafloor.

Drilling couldn't resume until the drilling contractor replaced the entire marine-riser system, which would take six to eight weeks. The first well was within twenty feet of the final planned depth of the well when the marine-riser system failed. Our primary and secondary reservoir objectives had minor, uneconomic quantities of oil. The first exploration prospect was a dry hole, and I had only one remaining ready-to-drill prospect in our inventory. The failure of the marine-riser system provided our team the necessary time to finalize the last two ready-to-drill exploration prospects for our deepwater drilling program. The marine-riser system failure would turn out to be a blessing for our deepwater exploration program.

Raji, one of the four geophysicists on the deepwater team, came to me with an exploration concept, or lead. The lead needed significant work to develop into a ready-to-drill prospect, but it was obvious the exploration lead had potential. The only problem was the lead wasn't "exactly" located within our exploration license. The southwestern border of our license not been defined by PETRONAS, because it was near a disputed area with the Sultanate of Brunei. However, if the license line was extended out a few miles, the lead would be well within our exploration license.

Additional technical work confirmed the lead was very prospective. After several conference calls with the vice president of international, our GM met with PETRONAS to discuss the ambiguous southwestern border of our license. PETRONAS agreed the area of interest would be included in Murphy Oil's exploration license if Murphy Oil would drill the prospect. Our GM's negotiating skills would prove time and time again to be invaluable to Murphy Oil's Malaysian operation.

In June 2002, the marine-riser system was replaced, and the semisubmersible rig commenced drilling the second exploration well. This prospect would test an entirely new exploration play and was located approximately one hundred miles southwest of the first exploration prospect, which was a dry hole.

The drilling of the second deepwater exploration well also went very well. The well encountered oil; however, the reservoir quality of the sandstones was very poor. After thorough analysis, the second well was classified as a dry hole. However, I was encouraged that the second well had found moveable oil, unlike the first exploration well.

Murphy Oil had the option to terminate the drilling contract due to the failure of the marine-riser system. Our CEO requested a conference call with the Malaysian office, the president of upstream, and the vice president of international to discuss whether to drill a third well.

The GM told me that our CEO liked to hear investment recommendations from "the horse's mouth." The conference call began at 8 p.m. on a Sunday in Kuala Lumpur, Malaysia. The GM, the drilling manager, and I were the only attendees for the call in Kuala Lumpur. The CEO had a series of questions on the reserve potential and the geological risks of the prospect. I spent approximately two hours answering his questions on the prospect and the remaining exploration potential of the deep-water licenses. After the two-hour discussion, our CEO said, "Let's drill this well."

The semisubmersible rig commenced drilling the third deepwater exploration well in Sabah, Malaysia. Raji, the geophysicist who first identified the lead, named the prospect, "Kikeh." I was told that Kikeh was a magical fish from Malaysian nursery rhymes. The Kikeh Prospect turned out to be magical for Murphy Oil and Malaysia.

Our CEO flew to Kuala Lumpur while we were drilling the Kikeh Prospect. He was flying to Kuala Lumpur to sign two new exploration licenses in the offshore region of Peninsular Malaysia. Our CEO would be arriving in Malaysia as we drilled through the reservoir objective of the Kikeh Prospect.

I would update him on status of the Kikeh Prospect and the exploration potential of the two deepwater Sabah leases and the two offshore Sarawak leases. Our CEO's timing meant I had to develop two different deepwater Sabah presentations, a Kikeh discovery and a Kikeh dry hole. If Kikeh was a dry hole, I would recommend the company stop drilling in the deepwater lease. I also knew our vice president of international wanted to drill a fourth exploration well in the deepwater lease. I thought the prospects of me getting fired if Kikeh were a dry hole were very high. However, "this horse" was going to give our CEO his honest opinion.

Logging-while-drilling (LWD) tools were used on all the deepwater Sabah exploration wells. LWD tools can measure the rock properties while the well is drilling. As we drilled into the objective section, the LWD tools indicated reservoir quality sands with hydrocarbons. Our operations geologist called me at 5 a.m., giving me a detailed report on the well. Our operations geologist thought we had oil, but we couldn't be sure until we acquired additional data. I immediately called our vice president of international, who I could tell was relieved to hear any encouraging news.

At 9 a.m., our GM, the drilling manager, and I gave our daily telephone report to our vice president of international. The drilling manager was eager to drill ahead, even though we had encountered hydrocarbon-bearing sands. I said, "I estimate we have approximately five hundred feet of hydrocarbons in five discrete sand packages. We don't know the formation pressures and we don't want to take a risk of losing this well." The vice president of international categorically agreed with me, much to the chagrin of the drilling manager.

We stopped drilling and immediately acquired additional wireline data, which confirmed that we indeed had five hundred feet of oil pay in high-quality sands. The data also confirmed that the formation pressures were significantly higher than we initially thought. If we had drilled ahead, as the drilling manager recommended, the well would have been put at risk.

The CEO arrived as we were completing the data acquisition of the Kikeh exploration well. I met with the CEO, the president of upstream, the vice president of international, and the GM for dinner that night. I brought with me small flasks with oil samples from the Kikeh exploration well. The oil was a light-yellow color and of a very high quality. All the data indicated we had a significant commercial discovery.

We had a very enjoyable dinner discussing the Kikeh discovery. Although, the vice president of international pulled me aside and said, "Next time, make sure the first exploration well is a discovery, not the third exploration well." Our GM told me our vice president's comment to me wasn't a joke. Obviously, I wasn't the only person feeling the pressure from our exploration programs.

Our deepwater Sabah drilling program immediately shifted from exploration drilling to appraisal drilling in the Kikeh Field. The results of the drilling program went very well, and the initial results indicated the Kikeh Field had ultimate estimated reserves (EUR) potential of four hundred million barrels of oil.

PETRONAS was ecstatic about the news of the Kikeh discovery. Our GM received a well-deserved standing ovation at a PETRONAS conference over Murphy Oil's success in deepwater Sabah, Malaysia. Our partner PETRONAS Carigali was also ecstatic about the news.

PETERONAS Carigali began to ask me what Murphy Oil saw that Royal Dutch Shell didn't see. Royal Dutch Shell's reputation was starting to tarnish in the eyes of PETRONAS and PETRONAS Carigali.

PETRONAS held a conference on exploration a few months after the Kikeh discovery. All the exploration managers from companies operating in Malaysia participated in a round-table discussion on technology. The PETRONAS moderator asked each exploration manager to speak on their experiences with different types of new technology.

Eight exploration managers participated in the round-table discussion. The first round-table discussion topic was logging-while-drilling (LWD). The ExxonMobil exploration manager gave a succinct summary of his experiences in the use and application of the LWD technology. I followed with my experiences from around the world and concluded with my recent experiences in the Kikeh discovery. The next person to speak was the exploration manager from Royal Dutch Shell, who sat in silence for three to four minutes. The exploration manager next to him then spoke up and gave his summary of the technology.

I leaned over to the Royal Dutch Shell exploration manager and said, "I am sure we can get the moderator to come back and let you give your assessment of LWD technology." He said, "That is all right; you see, I have never drilled a well of any type in my twenty-year career with Royal Dutch Shell." I was shocked that Royal Dutch Shell would put such an inexperienced person in a region with such high exploration activity. His inexperience would cause Royal Dutch Shell to lose significant value in Malaysia.

In 1994, Royal Dutch Shell drilled a major gas discovery, which was called the Kebabangan Field.[32] A significant accumulation of shallow gas covered the crest of

---

32   abarrelfull.wikidot.com

the structure. Royal Dutch Shell drilled on the flank of the large structure to avoid a potential drilling hazard. A second well on the flank of the structure encountered significant natural gas with a small quantity of oil.

Royal Dutch Shell elected not to drill any further appraisal wells in the 1990s on the Kebabangan Field. In Malaysia, if a company drills an oil discovery, the company has five years to commit to a development plan. If a company drills a gas discovery, the company has ten years to commit to a development plan.

Royal Dutch Shell assumed the oil encountered in the second well was insufficient to be economic and declared Kebabangan to be a gas field. The gas field declaration "gave away" Royal Dutch Shell's rights on future oil reserves in the Kebabangan Field. In 2002, Royal Dutch Shell drilled a third well to appraise the gas reserves of the Kebabangan Field. The third well was drilled on the structural crest of the structure and encountered hundreds of feet of oil-bearing sands.

PETRONAS notified Royal Dutch Shell that PETRONAS Carigali would be taking over the drilling operation of this well. Royal Dutch Shell had renounced all oil rights to the Kebabangan Field and any oil in the field belonged to the Malaysian government. A cry of anguish could be heard all the way back to Royal Dutch Shell's headquarters in the Netherlands.

The Kebabangan well results prompted several companies to submit proposals to PETRONAS to appraise and develop oil reserves in the Kebabangan Field. Murphy Oil submitted a proposal, but were told "we had been too successful, and it was time to bring in new company."

PETRONAS elected to form the Kebabangan Petroleum Operating Company (KPOC) to develop the field. The companies in KPOC are PETRONAS Carigali (40% equity), ConocoPhillips (30% equity) and Royal Dutch Shell (30% equity). I am certain this was not Royal Dutch Shell's desired outcome when the third well was drilled in the Kebabangan Field in 2002.

Barbara and I saw several global events that reshaped the world during our time in Malaysia. In March 2003, the United States invaded Iraq, starting the Iraq War. In the United States, several pacifists groups protested the United States' invasion of Iraq. A few days after the start of the Iraq War, there was a peaceful demonstration of approximately forty people outside the United States Embassy, protesting the United States invasion of Iraq. The Malaysian police outnumbered the demonstrators and the Malaysian military outnumbered the Malaysian police. It was a peaceful demonstration without any indications of violence.

That night, Barbara's parents called to see if we were going to be evacuated from Kuala Lumpur. Malaysia's Prime Minister Mahathir had spoken out against the United States invasion of Iraq. An American news agency had reported rioting outside the United States Embassy in Kuala Lumpur. We reassured Barbara's parents that Malaysia was a peaceful and law-abiding country. I suspect an overzealous news agency, feeling the pressure to report significant news twenty-four hours a day, seven days a week, grossly exaggerated the demonstrations in Kuala Lumpur, Malaysia.

Today, we see frequent news reports of the People's Republic of China (PRC) constructing military bases on tiny islands in the South China Sea. PRC's territorial claim to these islands is baseless and has been formerly rejected by the United Nations. PRC's baseless territorial claims aside, the South China Sea has a long-standing history of overlapping claims or territorial disputes, as shown in *Figure 23*.[33]

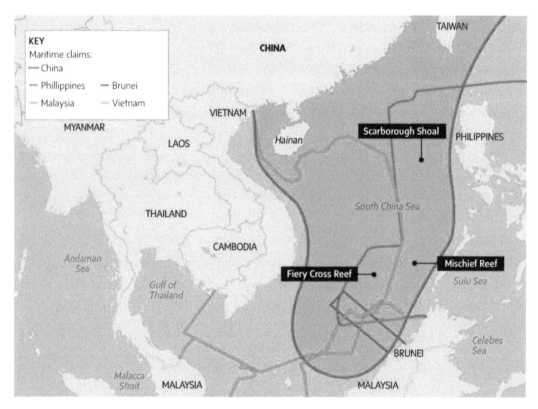

Figure 23

The Federation of Malaya became an independent country in 1957. In 1963, the Malaya Federation was enlarged by the accession of the Sarawak and Sabah. In 1963, Royal Dutch Shell discovered a giant gas field in Brunei, which prompted plans for

---

33   *Daily Mail,* May 22, 2015

an LNG plant. In 1963, Brunei elected to remain a British dependency, rather than join the Malaya Federation, which became known as Malaysia. Although Malaysia and Brunei have maintained close government-to-government relations, a dispute between the two countries was rapidly escalating over the deepwater region of northwest Borneo in 2003.

Malaysia claimed Brunei's maritime territory ended at a water depth of six hundred feet (100 fathom isobath), as stated in Order in Council 1958, Number 1518. Brunei claimed a maritime territory of two hundred nautical miles from its shoreline. Brunei's claim extended into the deepwater region of the South China Sea.

In 2000, Brunei held a lease sale for the disputed deepwater region of northwest Borneo. In 2002, Royal Dutch Shell, Mitsubishi, and ConocoPhillips submitted the most significant work program for the western deepwater lease and Total, BHP Billiton, and the Hess Corporation submitted the most significant work program for the eastern deepwater lease. However, the government of Brunei had yet to award either of the two deepwater exploration leases.

PETRONAS was uncertain how to respond to the pending deepwater license awards by the Brunei National Petroleum Company. Our GM provided PETRONAS a simple solution: Malaysia should preempt Brunei and make their own deepwater license awards for the disputed deepwater region. Our GM also recommended that PETRONAS award the licenses to a company that had absolutely no oil and gas interests in Brunei, such as Murphy Oil.

A frenzy of activity occurred within PETRONAS, which resulted in Murphy Oil and PETRONAS Carigali being awarded both deepwater leases in the disputed area with Brunei. The newspapers in Malaysia broke the news, much to the chagrin of the Sultanate of Brunei, Royal Dutch Shell, Mitsubishi, ConocoPhillips, Total, BHP Billiton, and the Hess Corporation.

I received a frantic call from the exploration manager with Royal Dutch Shell Brunei, asking me if Murphy Oil had been awarded the two deepwater leases in Brunei. I told him that the area was considered Malaysian maritime territory by PETRONAS. I reminded him that Royal Dutch Shell had an office in the same office building as PETRONAS in Kuala Lumpur. I suggested he call the exploration manager with Royal Dutch Shell Malaysia, as I was sure he would have a thorough understanding of the situation. Apparently, the exploration manager of Royal Dutch Shell Brunei did call their office in Kuala Lumpur, and his call was not well received. PETRONAS had kept the license award process a secret until it was announced in the newspapers.

Murphy Oil and their partner, PETRONAS Carigali had been awarded Block L and Block M in the Malaysian maritime territory. In a few weeks, the Brunei National Petroleum Company awarded Block J to Total, BHP Billiton, and the Hess Corporation and Block K to Royal Dutch Shell, Mitsubishi, and ConocoPhillips, as shown on *Figure 41*. Southeast Asia watched to see what would happen next.

Murphy Oil was prepared with a ready-to-drill prospect when PETRONAS announced the award of the two deepwater leases. Murphy Oil also had four deepwater rigs under contract that could drill the ready-to-drill prospect as soon as we received approval from the Malaysian government. Our drilling manager coordinated all our drilling operations very closely with the Malaysian government, Malaysian military, and PETRONAS.

The Malaysian government wanted to drill Murphy Oil's prospect in Block L as quickly as possible. In March 2003, Murphy Oil drilled the exploration prospect with Malaysian military naval and air support. The exploration well was an oil discovery and the Malaysian government directed Murphy Oil to move the drilling rig out of Block L. Malaysia wanted to initiate government-to-government negotiations with Brunei.

I heard rumors that Total was planning to drill an exploration well in their Block J, which was Murphy Oil's Block L. However, Total never drilled their deepwater exploration well. A few months later, an old friend who lived in Brunei told me that Total had mobilized a drillship to drill an exploration well in Block J. However, the Brunei military naval and air support never materialized, and the drillship was released.

On October 30, 2003, Dr. Mahathir Mohamad stepped down as prime minister of Malaysia and was succeeded by Deputy Prime Minister Abdullah Ahmad Badawi.[34] Malaysia had grown and prospered under Dr. Mahathir Mohamad's leadership for twenty-two years. The leadership of Prime Minister Badawi would include a disturbing footnote for Murphy Oil on the Malaysia-Brunei deepwater overlapping claim dispute.

On March 16, 2009, Prime Minister Badawi signed the "2009 Exchange of Letters" with Sultan Hassanal Bolkiah of Brunei.[35] Details of the agreement were not made public. On April 23, 2010, the Brunei government announced their country had "retained ownership" of the two petroleum blocks which Malaysia had previously claimed because of the 2009 Exchange of Letters.[36]

Dr. Mahathir Mohamad accused Prime Minister Badawi of "signing away"

---

34 CNN.com/WORLD, October 30, 2003
35 Dictionary of Modern Politics of Southeast Asia by J. Liow and M. Leifer, 2014
36 Durham University Centre for Borders Research, 2009

Malaysia's rights over hydrocarbon resources in the area, specifically in Blocks L and M, in exchange for Brunei giving up its claim over an onshore region in Sabah, Malaysia. Murphy Oil announced that Block L and Block M production, sharing a contract with PETRONAS and situated within Brunei's maritime territorial claim, had been terminated because they were "no longer a part of Malaysia." The secrecy and the result of the 2009 Exchange of Letters was not well received by the Malaysian public.

In 2003, Murphy Oil's corporate focus was on Malaysia. The development of the West Patricia Field was successfully completed, and the field was producing over fifty thousand barrels of oil per day. The appraisal of the Kikeh Field had also been successfully completed, and the company was rapidly moving forward with the field development and first oil production.

The Kikeh Field lies in a water depth of approximately four thousand three hundred feet and would be the first deepwater development in Malaysia. The development plan incorporated a Spar, a subsea water injection and production system, that would be connected to a Floating Production Storage and Offloading (FPSO) vessel. Murphy Oil's goal was to achieve first production within five years—an exceptional achievement.

Proven reserves are the foundation of every oil and gas company in the world. The financial community measures a company's performance on their ability to cost effectively replace their proven reserves. Companies can borrow money from banks based on their proven reserves. Our CEO was adamant that our company obtain an accurate, even conservative reserve assessment for the Kikeh Field. An outside reserve audit company was brought in to certify the proven reserves in the Kikeh Field. The reserve audit company stated the Kikeh Field was a textbook example of how to appraise a field and remove uncertainty in the reserve assessment. The world would soon see that not all companies adhered to strict reserve audit procedures.

On January 9, 2004, Royal Dutch Shell stunned the financial world by reducing its proven oil and gas reserves by 3.9 billion barrels. The reserve write-down represented an estimated one hundred thirty-five billion-dollar reduction in Royal Dutch Shell's future revenue. Royal Dutch Shell's shares plunged, and the chairman, Sir Philip Watts, and the head of exploration and production, Mr. Walter van de Vijver, resigned over Royal Dutch Shell's shocking reserve announcement.

The financial community couldn't understand how a twenty percent reserve reduction could happen to a company like Royal Dutch Shell. However, the book, *Shell Shock: The Secrets and Spin of an Oil Giant,* by Ian Cummins and John Beasant, provides insightful analysis of the company, including questionable reserve reporting procedures. The book also explains how someone who had no exploration operations experience could become the exploration manager for Royal Dutch Shell in Malaysia.

Approximately six miles south of the Kikeh Field was the Kakap Prospect. The Kakap Prospect looked like a direct analogue to the Kikeh Field. However, the 3-D seismic quality over the Kakap Prospect was very poor due to a massive accumulation of shallow gas. The shallow gas would make drilling the prospect challenging. The Kakap Prospect also extended into another Malaysian deepwater lease, which was operated by Royal Dutch Shell.

Murphy Oil and Royal Dutch Shell both drilled significant oil discoveries on the same structure in our respective blocks. Royal Dutch Shell called their discovery the Gumusut Field and Murphy Oil called our discovery the Kakap Field. It was now up to PETRONAS to determine who would operate the Gumusut-Kakap appraisal and development program.

PETRONAS organized a series of technical meetings to review the subsurface potential of this field. Royal Dutch Shell brought in six technical experts from the Netherlands to demonstrate their technical superiority. Surprisingly, none of the Royal Dutch Shell technical experts had any deepwater experience and weren't even familiar with the Gumusut-Kakap exploration well results.

In 2004, Murphy Oil had drilled more wells in deepwater Sabah, Malaysia, than Royal Dutch Shell. The subsurface characteristics in the Kikeh Field and the Gumusut-Kakap Field were virtually identical. Murphy Oil had a significantly greater subsurface knowledge and expertise of the Gumusut-Kakap Field than Royal Dutch Shell. This knowledge and expertise would save significant time and money in the appraisal and development drilling of this field.

Subsurface expertise aside, hundreds of millions of dollars could be saved by integrating Murphy Oil's Kikeh Field facilities with the Gumusut-Kakap Field facilities. The Kikeh Field was on schedule to be fully developed and on production within five years. Royal Dutch Shell's plans didn't seek to take advantage of the Kikeh Field facilities and was forecast to take at least ten years to achieve first oil production.

Selecting Murphy Oil as operator of the Gumusut-Kakap Field appraisal and development program offered PETRONAS and the country of Malaysia the greatest financial return. However, I could tell that the PETRONAS senior manager, who would select the Gumusut-Kakap Field operator, didn't want Murphy Oil to be the operator.

PETRONAS Carigali was a partner in Royal Dutch Shell's block and Murphy Oil's block that contained the Gumusut-Kakap Field. I went to the PETRONAS Carigali exploration manager and asked him if his company would consider being the operator of this field's appraisal and development program. He told me PETRONAS Carigali desperately wanted to gain deepwater appraisal and development expertise and experience, but he didn't have the sufficient staff for such a large project. I suggested that Murphy Oil and almost certainly Royal Dutch Shell could second technical experts to the project who could fill PETRONAS Carigali's technical skill gaps and provide training to his staff. The PETRONAS Carigali exploration manager thought this was a brilliant idea and said he would raise the idea with PETRONAS senior management.

Two weeks later, the PETRONAS Carigali exploration manager told me he couldn't talk about the idea we had previously discussed. He then told me Royal Dutch Shell would be selected to be the operator of the Gumusut-Kakap Field appraisal and development program. He implied that Royal Dutch Shell had granted the PETRONAS senior manager "special privileges" to become the operator of the Gumusut-Kakap Field.

Royal Dutch Shell was awarded operatorship of the Gumusut-Kakap Field. It took Royal Dutch Shell twelve years to achieve first oil production in this field after the initial exploration well. Why did the PETRONAS senior manager select Royal Dutch Shell to be the operator of the Gumusut-Kakap Field? One can only speculate.

In 2004, PETRONAS was concerned about the long-term supply of natural gas to their Bintulu LNG complex in Sarawak. Long-term supply of natural gas is a critical factor to LNG buyers, particularly major importers like South Korea and Japan. The potential natural gas supply shortage created an opportunity for my company in Malaysia.

Royal Dutch Shell had been the previous operator of Murphy Oil's two shallow water blocks in Sarawak. Royal Dutch Shell drilled numerous exploration wells in the two blocks, many of which encountered natural gas. Royal Dutch Shell believed the natural gas reserves were too small to be economic and released the two blocks.

Murphy Oil acquired 3-D seismic data over the two shallow water blocks and recognized the oil and gas potential that had been missed by Royal Dutch Shell. The Sarawak shallow water exploration program was a resounding success. In September 2009, Murphy Oil began producing natural gas to the PETRONAS Bintulu LNG Complex in Sarawak.

In 2004, I received a mysterious call from PETRONAS to come to their office for a meeting. I was told the purpose of the meeting would be explained to me once I arrived at the PETRONAS office. When I arrived, the PETRONAS secretary took me to a small windowless office. A PETRONAS manger came into the office, introduced himself, and asked me to take fifteen minutes to read a document. The document was a shallow hazard drilling assessment for a Royal Dutch Shell exploration well. The PETRONAS manager then left me to read the document.

In fifteen minutes, the PETRONAS manager returned and asked me for my thoughts on the document. I said, "The report states there are no shallow hazards, yet the seismic data in the report clearly shows two distinct shallow hazards." I then described the potential outcome of drilling through each of the two potential drilling hazards. The PETRONAS manager said, "Congratulations, you have just described what happened on this recent Royal Dutch Shell well."

Apparently, Royal Dutch Shell had encountered an uncontrolled flow of hydro-carbons to the surface on three consecutive deepwater wells. For the next two years I was called in to secretly review Royal Dutch Shell's shallow hazard assessments for their deepwater exploration wells. PETRONAS no longer had the illusion that Royal Dutch Shell was infallible.

PETRONAS was constantly asking me, "How can a small, independent company like Murphy Oil consistently outperform a major company like Royal Dutch Shell?" I decided to evaluate how the oil and gas industry had evolved over the previous decades. I thought this analysis would provide answers to PETRONAS's question.

As I did my evaluation, I saw several trends that surprised even me. My presenta-tion of the analysis was well received by PETRONAS. I was also asked to give this presentation at a major energy conference in Den Hague, Netherlands. This presen-tation was again well received, especially by several representatives from OPEC.

Dr. Peter Rose, president of the American Association of Petroleum Geologists (AAPG), asked me to give my presentation at the International American Association of Petroleum Geologists in Perth, Australia. Dr. Rose thought the industry needed to be shaken up and he thought my presentation might open some eyes.

My presentation was entitled, "Hydrocarbon Exploration and Production: The Evolution and Revolution of an Industry." I made the following five key points in my presentation:

1. From the 1880s to 1980s, large multinational oil companies like Exxon, Royale Dutch Shell, and British Petroleum dominated all facets of the oil industry. For one hundred years, large multinational oil companies led in the development of new technology, exploration for new hydrocarbon reserves, oil production, and refining.

2. In the 1980s, global oil prices declined dramatically. Large multinational oil companies now struggled to remain financially solvent. Companies cut or even eliminated research and development. Overhead was slashed, resulting in massive layoffs across the industry. Exploration for new hydrocarbon reserves was no longer important, as the world was awash in oil. Acquisitions and mergers usually occur when any industry experiences a downturn and the oil industry was no different. From 1982 to 2006, large, multinational companies spent over three hundred billion dollars in company and major asset acquisitions, as shown in *Figure 24*.

	ACQUISITIONS	PRICE	1982	1985	1990	1995	2000	2005
1	OXY ACQUIRES CITY SERVICES	$3.5 Billion	▮					
2	TEXACO ACQUIRES GETTY	$10.4 Billion	▮					
3	MOBIL ACQUIRES SUPERIOR	$5.7 Billion	▮					
4	CHEVRON ACQUIRES GULF	$13.2 Billion	▮					
5	BP ACQUIRES BRITOIL	~$4 Billion			▮			
6	TENNECO SELLS ALL ASSETS	$7.3 Billion			▮			
7	BP ACQUIRES AMOCO	$48.2 Billion					▮	
8	TOTAL ACQUIRES FINA	$12.9 Billion					▮	
9	BP ACQUIRES ARCO	$26.8 Billion					▮	
10	TOTAL ACQUIRES ELF	~$11 Billion					▮	
11	REPSOL ACQUIRES YPF	$13.4 Billion					▮	
12	EXXON ACQUIRES MOBIL	$81.0 Billion					▮	
13	BP ACQUIRES BURMAH CASTROL	$4.7 Billion						▮
14	CHEVRON ACQUIRES TEXACO	$38.7 Billion						▮
15	PHILLIPS ACQUIRES CONOCO	$15.2 Billion						▮
16	CHEVRON ACQUIRES UNOCAL	$18.3 Billion						▮

TOTAL ESTIMATED PRICE ~ $ 314 BILLION

Figure 24

Many companies were faced with a bleak financial future due to high operating cost and corporate debt. These companies actively sought to be acquired. The acquiring companies saw an opportunity to acquire new reserves without the risk of exploration.

3. As the large multinational companies were shrinking, the public sector, or national oil companies, were rapidly expanding across the world. The growth of the national oil companies occurred in both oil-importing and oil-exporting countries. The oil-importing countries, like Japan, Korea, and Italy, formed national oil companies to establish a secure supply of oil. The oil-exporting countries, like Indonesia, Norway, Nigeria, and Venezuela formed national oil companies to maximize the value of their oil resources. The contraction of the large multinational oil companies resulted in the national oil companies becoming the dominant oil producers in the world. *Figure 25* compares the top ten oil-producing companies in 1972 and in 2006.

TOP TEN OIL PRODUCERS

	1972			2006		
	COMPANY NAME	TYPE COMPANY	AVERAGE DAILY NET PRODUCTION MBO / DAY	COMPANY NAME	TYPE COMPANY	AVERAGE DAILY NET PRODUCTION MBO / DAY
1.	EXXON	Major	4,968	ARAMCO	NOC-Saudi Arabia	11,663
2.	BP	Major	4,664	GAZPROM	NOC-Russia	8,874
3.	SHELL	Major	4,169	NIOC	NOC-Iran	6,036
4.	TEXACO	Major	3,777	PEMEX	NOC-Mexico	4,328
5.	CHEVRON	Major	3,232	EXXON MOBIL	Major	4,237
6.	GULF	Major	3,214	BP	Major	3,878
7.	MOBIL	Major	2,316	CNPC	NOC-China	3,478
8.	ROSNEFT	NOC-USSR	1,301	SHELL	Major	3,425
9.	TOTAL	Major	977	SONATRACH	NOC-Algeria	3,172
10.	SONATRACH	NOC-Algeria	925	PDVSA	NOC-Venezuela	2,895

Figure 25

In 2006, ARAMCO, the Saudi Arabian national oil company's daily oil production was more than twice ExxonMobil's daily oil production. In 1972, only seven percent of the total daily oil production from the top ten oil-producing companies was from national oil companies, as shown in *Figure 26*. In 2006, 78% of the total daily oil production from the top ten oil producing companies was from national oil companies. In twenty-five years, the national oil companies had become the dominate oil producers in the world.

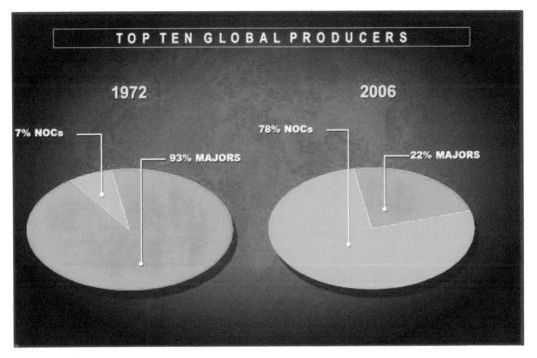

Figure 26

4. The service companies, like Schlumberger and Halliburton, began to invest heavily in new technology development. In 2006, the four largest service companies (Schlumberger, Halliburton, Baker Hughes, and Weatherford) were awarded over eight hundred exploration and production patents. The four largest private sector majors (ExxonMobil, British Petroleum, Total, and Royal Dutch Shell) were awarded approximately one hundred exploration and production patents.

5. Large multinational oil companies dramatically cut all nonessential expenditures as the global price of oil declined. High-risk and high-reward exploration projects were dramatically cutting the budgets of companies. Large multinational companies never regained the appetite for exploration of new reserves after the 1980s, as shown in *Figure 27*.

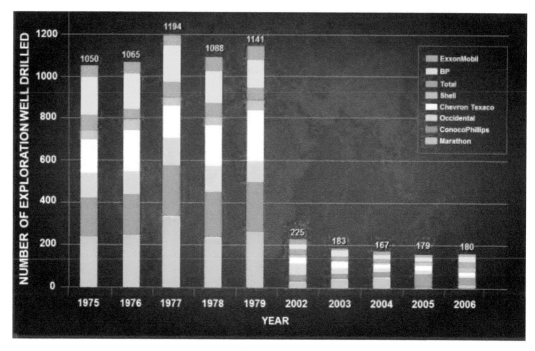

Figure 27

The small, dynamic oil companies quickly seized the international exploration opportunities abandoned by many of the large multinational oil companies.

Approximately 36,000 exploration wells were drilled across the world from 2002 to 2006. Small, dynamic oil companies drilled 87%, national oil companies drilled 7%, and large multinational oil companies drilled 6% of the exploration wells across the world over this period. Small, dynamic oil companies now dominate global exploration, onshore, offshore, and even in deepwater environments. The large multinational oil companies have become risk averse.

Independents now dominate global exploration, including in deepwater environments. To me, the statistics indicate the major companies had become risk averse. National oil companies dominate global oil production and service companies dominate the development of new technology. Have large multinational oil companies become irrelevant?

My presentation was well received at the International American Association of Petroleum Geologists. Most of the comments I received were of shock at the precipitous drop in exploration drilling by the large multinational companies. However, one person from ExxonMobil told me the important message from my talk was ExxonMobil needed to be more diligent in applying for patents on new exploration and production technology. When I heard this comment, I knew I had made the correct decision not to stay when Mobil was acquired by Exxon.

I received a call from Murphy's vice president of international while I was in Perth, Australia. The vice president said he was being promoted to president of Murphy Exploration and Production. He asked me to become the vice president of exploration for the Gulf of Mexico. Barbara and I discussed the opportunity, and I accepted the position.

Murphy Oil's market capitalization had grown from $2.9 billion in June 2001 to $9 billion in June 2006, primarily due to Murphy Oil's success in Malaysia.[37] I thought I could add greater value to Murphy Oil's Gulf of Mexico exploration program, than I could add value if I stayed in Malaysia. However, I would once again be going from the frying pan into the fire.

In 2006, as Barbara and I were preparing to move from Kuala Lumpur, Malaysia to Houston, Texas, the average price for oil in the United States was $58.30 per barrel. From 2001 to 2006, the price of oil had increased by $35.30 per barrel. The United States was producing 5.09 Million Barrels of Oil per Day (MMBOPD), while consuming 20.69 MMBOPD a day. From 2001 to 2006, the amount of oil produced in the United States decreased by over 700,000 Barrels of Oil per Day (BOPD). From 2001 to 2006, the consumption of oil in the United States increased by more than 1,000,000 BOPD,[38] as shown by *Figure 28*.

37   Quandl.com/topics/mur-market-cap
38   U.S. Energy Information Administration – Total Energy Review

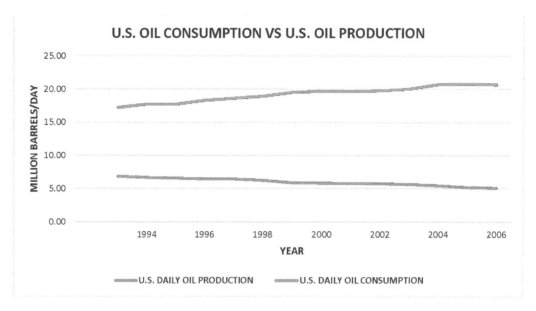

Figure 28

One of the reasons for the dramatic oil price increase from 2001 to 2006 was the soaring demand for oil in People's Republic of China (PRC). In the PRC, consumption of oil far exceeded the domestic supply of oil, as shown in *Figure* 29.[39] Once again, the global demand for oil had rapidly exceeded supply and the world economies were not prepared.

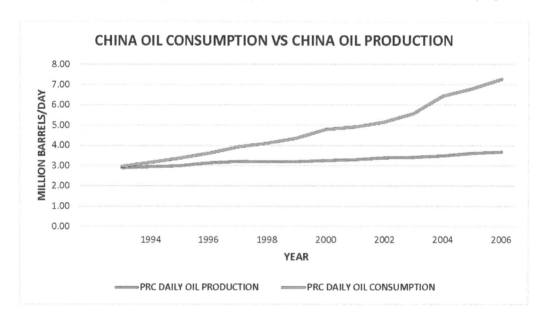

Figure 29

39   U.S. Energy Information Administration

Global fuel subsidies also contributed to escalating demand for oil. The Economist reported: "Half of the world's population enjoys fuel subsidies. This estimate, from Morgan Stanley, implies that almost a quarter of the world's petrol is sold at less than the market price." U.S. Secretary of Energy, Samuel Bodman, stated that approximately thirty million barrels per day of oil consumption was subsidized. Subsidized oil prices will never encourage energy conservation.

In 2006, the average price for gasoline in the United States was $2.59 per gallon, which is equivalent to $3 per gallon, when the price is adjusted to inflation (March 2015). From 2001 to 2006, the price of gasoline in the United States had increased $1.13 cents per gallon. The one dollar and thirteen cent price increase in gasoline prices from 2001 to 2006 was still not an incentive for the American public to conserve energy.

In 2006, the coal industry in the United States was still experiencing production growth, as shown in *Figure 30*.[40] The global demand for coal continued to grow due to the low price and abundant supply.

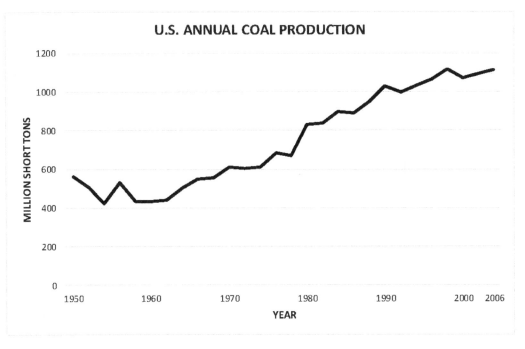

Figure 30

In the United States, escalating oil and gas prices and growing concerns about security of energy supply prompted renewed interest in renewable energy. In August

40   U.S. Energy Information Administration – Total Energy Review

2005, President George W. Bush signed the Energy Policy Act, which authorized tax credits for wind and other alternative energy such as wave and tidal power. This legislation triggered a five-fold increase in wind energy production in the United States, as shown in *Figure 31*.[41] I believe 2005 was the start of the wind energy revolution in the United States. It would take another nine to ten years before solar energy would experience a similar growth rate.

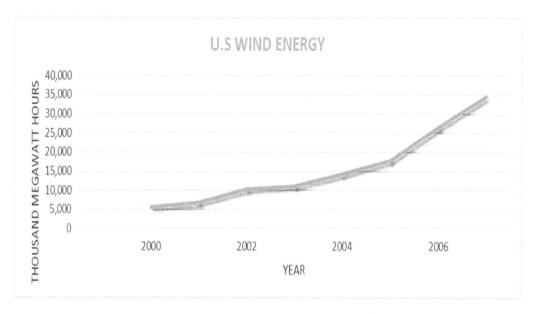

Figure 31

41   American Wind Energy Association

CHAPTER 15

Righting A Sinking Ship

Barbara and I moved from Kuala Lumpur, Malaysia, to Houston, Texas, in November 2006. Murphy Oil's office for the Gulf of Mexico was in New Orleans, Louisiana. However, the city of New Orleans had been devastated by Hurricane Katrina in August 2005. More than a year after the hurricane, the city of New Orleans was struggling to offer even basic services for the residents.

Murphy Oil's new president of upstream's plan was to move the office from New Orleans to Houston. I would be based in Houston but would fly every week to New Orleans to lead the exploration program. Barbara and I found a nice town-house within six miles of our Houston office and twenty miles from George Bush International Airport. I knew I would be spending many long nights and weekends in our Houston office over the next few years.

When I was in New Orleans, I stayed at a nice furnished apartment a few miles from our downtown office. Crime was rampant in New Orleans after Hurricane Katrina. My apartment was surrounded by high barbed wire fences with armed security guards patrolling the complex twenty-four hours a day, seven days a week. Several nights I had flashbacks from my days in Vietnam.

Although the apartment was modern and well maintained, electrical power outages occurred almost daily. Most of the power outages occurred early in the morning, as I was getting up to go to work. I soon became proficient at shaving in the morning using the light from my BlackBerry phone. The light from my BlackBerry also helped me walk the pitch-black hall to my car.

The morale of the exploration team was as dark as the pitch-black hall in my apartment. Most of the New Orleans staff had elected to retire and were counting the days until they could leave Murphy Oil. When I first flew into New Orleans, I met with the team to review their entire exploration portfolio. After two days, I knew we had a prospect inventory of zero. I realized this would be one of the most difficult challenges I had ever faced.

Murphy Oil's Gulf of Mexico assets consisted of two small deepwater fields and a few offshore exploration leases with no ready-to-drill prospects. The information technology (IT) support was almost nonexistent and the digital database was in chaos. The only good news was I met a geophysical consultant working in our office who I knew from my days working at Mobil Oil in New Orleans. He was an excellent geophysicist, and he was willing to join the company and move to Houston. This was the start of our new Gulf of Mexico exploration team.

Murphy Oil's upstream president was very interested in the 2007 Gulf of Mexico offshore lease sale. The United States Department of the Interior subdivided the Gulf of Mexico into the Western, Central, and Eastern Planning Area, as shown in *Figure 32*.[42]

Figure 32[43]

The Gulf of Mexico Central Planning Area Sale Number 205 was going to be held on October 3, 2007. This sale was expected to be very competitive, due to recent major oil discoveries in the deepwater region of the Central Planning Area. An unusually large number of deepwater leases would also be available at Sale Number 205. The Department of the Interior anticipated the amount of total high bids would exceed one billion dollars.

Most companies form partnerships for lease sales. A partnership allows the

---

42   Bureau of Ocean Energy Management
43   U.S. Department of Interior BOEM Map

companies to pool their financial resources to bid on more leases. This strategy is essential if the lease sale is going to be very competitive. Murphy Oil's previous lease sale partnership had dissolved, because the two partner companies had been acquired by other companies.

After I returned to Houston, our CEO visited our office to discuss Sale Number 205. He told me the company had budgeted twenty-five million dollars for bids at the upcoming lease sale. Our CEO thought Murphy Oil could win four or five prospective leases for twenty-five million dollars. My initial thought was twenty-five million dollars might win one, possibly two leases in Sale 205.

Our president wanted to aggressively grow Murphy Oil's exploration and production position in the Gulf of Mexico. I could tell he expected our Gulf of Mexico operation to have the same type of success as our Malaysian operation. However, the competition in the Gulf of Mexico was significantly greater than Malaysia.

In Malaysia, three to four companies usually participate in a PETRONAS exploration license round. Over one hundred and fifty companies were expected to participate in Gulf of Mexico Sale 205. Competition in Gulf of Mexico lease sales is always intense. Most of the blocks draw multiple bids. Some blocks may receive bids from ten or more companies. The probability of overpaying for a lease, the aforementioned winner's curse, in Sale 205 was almost a certainty.

I met with our landman to discuss Sale 205. Our landman thought the total amount of high bids could exceed two billion dollars, which was more than twice the Department of the Interior's estimate. He said Murphy Oil might be able to join a bid group, but only as a nonoperator. We both knew that our president wanted Murphy Oil to be the operator of any Sale 205 bid group.

Our landman told me that several overseas companies had expressed interest in joining Murphy Oil in oil and gas ventures in the United States. I asked him to inquire if any of the companies wanted to join a Murphy Oil lease sale partnership for Gulf of Mexico Sale 205. I thought this would be our best opportunity to form a viable lease sale partnership as an operator.

In 2004, geophysical service companies like WesternGeco and Compagnie Générale de Géophysique (CGG) saw a commercial opportunity in Sale 205. The geophysical service companies began acquiring massive 3-D speculative seismic surveys in the Central Planning Area. The cost to license one of the massive speculative 3-D seismic surveys in Sale 205 ranged from twenty million dollars to forty million dollars.

A speculative or nonexclusive seismic survey is acquired and processed by a seismic contractor and then licensed for a fee on the open market. The seismic contractor owns the exclusive rights to the seismic data. A company pays a licensing fee to use the seismic data for a finite period of time. The cost for speculative seismic data is approximately forty percent of the cost of the actual acquisition price. The geophysical service companies hoped to sell their speculative 3-D seismic surveys dozens of times for Sale 205.

On my return trip to New Orleans, I met with the exploration team to discuss where to focus our exploration efforts with our limited resources. To pursue the "hot" deepwater oil play, Murphy Oil would need to spend approximately twenty million dollars to fifty million dollars to license the latest speculative 3-D seismic surveys.

The problem with this option was the major companies had purchased this data months ago and had an army of geophysicists currently working the 3-D seismic data. Our landman thought that each lease in the "hot" deepwater oil play could have bids of forty million dollars to seventy-five million dollars. At those prices, Murphy Oil wouldn't win even one lease.

Another drawback to the "hot" deepwater oil play was the depth of the wells. The depth of each of these wells would be twenty-five thousand feet to thirty thousand feet. The cost of one of these exploration wells would be between one hundred fifty million dollars to two hundred and fifty million dollars. The probability of commercial success in this play was only 30% . Financially, this wasn't a play that Murphy Oil wanted to pursue.

An alternative to the "hot" deepwater oil play was an exploration play in the eastern area of Central Planning Area. Many of these leases had never been leased, which was unique in the Gulf of Mexico. The reserve potential of the prospects in this area was much smaller than the prospects in the "hot" deepwater oil play. However, the well costs in this play were only fifteen million dollars to twenty-five million dollars and the probability of commercial success was forty to fifty percent. The economics of this play were robust and our landman expected less competition in this area than the "hot" deepwater oil play for Sale 205.

Murphy Oil purchased the only speculative 3-D seismic survey over our area of interest in Sale 205. I found out that Murphy Oil was only the fourth company that had purchased the speculative 3-D seismic data. I also found out that the geophysical service companies had sold their massive 3-D speculative seismic surveys in the "hot" deepwater oil play dozens of times. This indicated to me that we had the right strategy for Murphy Oil for Sale 205.

Murphy Oil finally closed the New Orleans office and relocated everyone to

Houston. I had been actively recruiting people and was fortunate to get eight very experienced technical professionals to join our Gulf of Mexico exploration team. I was beginning to think we had a real chance to succeed in Sale 205.

In March 2007, our landman told me that the Korean National Oil Company (KNOC) had expressed an interest in participating with Murphy Oil in Sale 205. KNOC is the national oil company of South Korea and is one of the most important industrial companies in the country. Over the next month, we had several technical meetings with KNOC. After these meetings, we didn't hear anything from KNOC for another month. I assumed KNOC had decided not to join our lease sale bid group.

In June, KNOC asked if we would present our exploration program to a team of consultants. We spent the month of June reviewing our technical work, leads, and prospects with the consultants. In my discussion with the consultants, I found out that KNOC had not had a commercial exploration well in over three years. KNOC was seeking low risk exploration opportunities. The consultants recommended KNOC join Murphy Oil's lease sale bid group.

As we were meeting with KNOC's consultant, our CEO brought in Stephens Producing Company to discuss joining Murphy Oil in Sale 205. It only took Stephens Producing Company a few weeks to review our work and to agree to join Murphy's bid group. After several months, KNOC also agreed to join Murphy's bid group. I was overjoyed when our landman told me Murphy had a bona fide bid group for Sale 205.

The cost for companies to join the Murphy Oil's bid group was steep. Murphy Oil had developed the lease sale strategy, purchased the speculative 3-D seismic licenses, and developed the exploration prospects. A partner would pay a "two-for-one promote" for every lease dollar bid on each block. As an example, if the lease sale partnership bid fifty million dollars for one lease, then KNOC would pay thirty-five million dollars (70%) for a 35% working interest in the lease. Stephens Producing Company also had the same lease sale partnership promote. After the lease was awarded, the partners would pay their normal working interest on all future costs, such as drilling an exploration well.

One month prior to Sale 205, our CEO visited our office to get an update on Sale 205. He was pleasantly surprised at the quality and number of prospects our exploration team had identified for the upcoming lease sale. I think he was even more impressed with the terms our partners had to pay to join our bid group. At the end of the meeting, he asked if we needed any more money to increase our bids for the lease sale. I thanked him and said, "No thank you, our bids are based on the risked net present value of each prospect, and we don't want to overpay for a lease." I could also tell our president was pleased with what we had been able to achieve.

On October 3, 2007, the U.S. Department of the Interior announced the results of Sale 205.[44] A total of eighty-four lease sale bid groups submitted bids totaling $5,245,583,944 on 723 blocks in the Central Planning Area! The total sum of high bids was $2,904,321,011. Sale 205 had been extremely competitive and many of the companies came away disappointed with the results.

Royal Dutch Shell submitted the single highest bid of $90,488,455 on a single block. The second highest bidder on this block bid approximately fifty million dollars. Royal Dutch Shell had the highest total of high bids at $554,563,223 on sixty-nine blocks. Chevron had the second highest total of high bids at $283,354,944 on forty-four blocks. I thought many of the high bidders would experience winner's curse from Sale 205. Over the next few years, drilling results proved me correct.

Large, independent companies like Anadarko Petroleum and Hess Corporation were successful in less than forty percent of their bids. These companies normally had a lease sale success rate of greater than seventy percent. Most of the independent companies in Sale 205 were very disappointed and reassessed their bid strategy for the next Gulf of Mexico lease sale.

The Murphy Oil bid group partnership spent $161,054,671 to win twenty-six blocks.[45] Murphy Oil spent less than twenty-five million dollars net, due to the bid group partnership promote. Murphy Oil won twenty-six of thirty-three blocks, which was a 79% success rate. Murphy Oil had competition on almost every block and was the highest bidder by the narrowest margin. In retrospect, all the pieces came together at just the right time. It wasn't genius, it was luck, pure and simple.

The local newspapers touted Murphy Oil as having an insightful strategy for Sale 205. Our landman told me that Murphy Oil was now getting phone calls to join multiple bid groups for Central Planning Area Sale 206. Ultimately, Sale 205 allowed Murphy Oil to reestablish itself in the Gulf of Mexico. After the sale, our president came up to me and said, "Excellent," which were the first and only positive words he had ever spoken to me.

The drilling manager in Malaysia was promoted to be vice president of operations in Malaysia. He did an excellent job of bringing the Kikeh Field on production in five years and on budget. In September 2007, he was promoted to senior vice president of North America. He would be my new boss.

The Gulf of Mexico Central Planning Area Sale Number 206 was going to be held

---

44   U.S. Department of the Interior Minerals Management Service Gulf of Mexico Region – October 3, 2007
45   Murphy Oil Corporation Form 10-K Report – February 29, 2009

on March 19, 2008. This sale was also expected to be very competitive, due to recent oil discoveries in the deepwater region of the Central Planning Area. The industry anticipated Sale 206 to be as competitive as Sale 205.

Murphy Oil's exploration team had identified two prospective areas in Sale 206. One area was an emerging oil play that I thought would have limited industry competition. The second area was in the "hot" deepwater oil play. The prospect of interest keyed off an offset oil discovery, drilled by Anadarko Petroleum.

Murphy Oil's landman approached Anadarko Petroleum and our two companies formed a bid group. The technical assessments of our two companies was very similar. Both companies saw one open block as an extension of a recent oil discovery, which had been drilled by Anadarko Petroleum. Murphy's success in Sale 205 was the primary reason Anadarko Petroleum agreed to the partnership for Sale 206.

The reserve potential of the prospect was significant, and the risk was low for exploration. Both companies recognized this block would require a substantial bid to have a chance to win the block at Sale 206. Anadarko recommended bringing in another company, Samson Offshore, into the bid group to provide additional financial resources for the bid on this block.

The U.S. Department of Interior has strict rules on how a lease sale partnership can develop their bids for each federal lease. The companies in a bid group sit around a table and the operator proposes a starting bid and a dollar increment, and each company can increase the bid. The operator starts the bidding, and then the next company can either increase the bid or pass.

The exploration team spent a significant amount of time looking at the reserve potential and geological risk of the prospective lease in the "hot" deepwater oil play. After the technical analysis, I spent many hours evaluating the economic potential of the prospect on the prospective lease. I thought the partnership should bid between seventy-five million dollars to eighty-five million dollars for this lease. However, the federal government forbids discussing oil price forecasts or sharing economic analysis in the lease sale bid meetings. The strict federal government policies are designed to encourage competition among the oil and gas companies. In my opinion, many of these federal policies are excessive.

At our lease sale bid meeting, it became apparent that Anadarko Petroleum and Samson Offshore wanted to bid significantly more money than Murphy Oil for this lease. I had to make several phone calls to our president and North American senior vice president to get approval to increase our bid about my approved limit. These phone calls were not well received by either the president or the senior vice president. After more than one hour, our bid group concluded the bid process with a bid of $105,600,789 for the one block. I walked out of the partner meeting thinking it would turn out to be a classic case of winner's curse.

On March 19, 2008, the U.S. Department of the Interior announced the results of Sale 206.[46] A total of eighty-five companies submitted bids totaling $5,740,047,263 on 615 blocks in the Central Planning Area. The total sum of high bids was $3,677,688,245. The total sum of high bids in Sale 206 was greater and for fewer blocks than Sale 205!

Anadarko Petroleum, Murphy Oil, and Samson Offshore submitted the single highest bid on a single block of $105,600,789 for Sale 206. The second highest bid on this block was approximately fifty-five million dollars. The Hess Corporation had the highest total of high bids of $437,541,152 on twenty-four blocks.

Murphy Oil had submitted a total of $37,890,211 on high bids on five blocks.[47] I had nightmares of drilling a dry hole on the $105,600,789 block. I knew our partnership had fallen into the winner's curse trap. Fortunately, the exploration well on the high-dollar block was a commercial discovery.

After Sale 206, our president asked me to lead the exploration program in Australia. Barbara was ready to pack as soon as I told her of the opportunity to live in Perth, Australia. We had visited this beautiful country many times on holiday. We also had good friends living in Perth. However, I would once again be going from the frying pan into the fire one last time with Murphy Oil.

In July 2008, as Barbara and I were preparing to move from Houston, Texas, to Perth, Australia, the average price for oil in the United States was $91.48 per barrel. From 2006 to 2008, the price of oil had increased by $33.18 per barrel. The United States was producing 5 Million Barrels of Oil per Day (MMBOPD), while consuming 19.50 MMBOPD. From 2006 to 2008, the amount of oil produced in the United States decreased by over 85,000 Barrels of Oil per Day (BOPD). From 2006 to 2008, the consumption of oil in the United States decreased by more than 1,100,000 BOPD,[48] as shown by *Figure 33*. The high oil price was once again encouraging oil conservation in the United States.

46   U.S. Department of the Interior Minerals Management Service Gulf of Mexico Region – March 19, 2008
47   Murphy Oil Corporation Form 10-K Report – February 29, 2009
48   U.S. Energy Information Administration – Total Energy Review

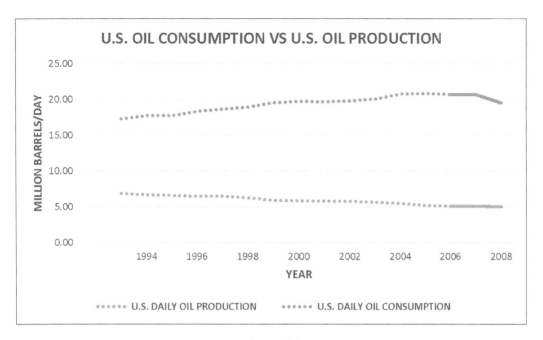

Figure 33

One reason for the dramatic oil price increase from 2006 to 2008 was the soaring demand for oil in People's Republic of China (PRC).[49] In the PRC, consumption of oil was rapidly exceeding the domestic supply of oil, as shown in *Figure 34*. Another reason for the exceptionally high oil prices was that many developing countries provide fuel subsidies. The fuel subsidies discouraged the conservation of oil.

---

49   U.S. Energy Information Administration – China May 2014

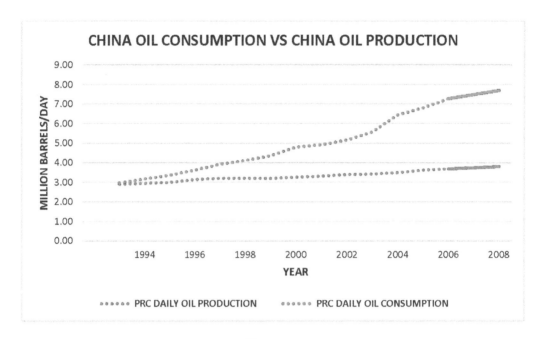

CHINA OIL CONSUMPTION VS CHINA OIL PRODUCTION

Figure 34

In 2008, the average price for gasoline in the United States was $3.27 per gallon, which is equivalent to $3.61 per gallon, when the price is adjusted to inflation (March 2015). From 2001 to 2008, the price of gasoline in the United States had increased $1.81 cents per gallon. The one dollar and eighty-one cent price increase in gasoline prices from 2001 to 2008 was beginning to cause the American public to conserve energy.

In 2008, the coal industry in the United States was nearing a historic production peak, as shown on the production peak in *Figure 35*.[50] The global demand for coal continued to grow due to the low price and abundant supply.

---

50   U.S. Energy Information Administration – Total Energy Review

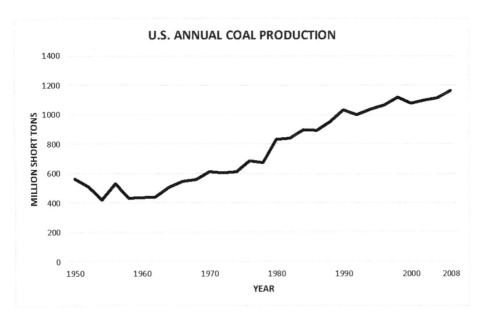

Figure 35

From 2001 to 2008, there was a negatable increase in electricity generated from nuclear power plants in the United States, as shown on *Figure 36*. Rising hydrocarbon prices and growing concerns over greenhouse gases gave the nuclear power industry hope of revival in the United States. However, renewed interest in nuclear power never materialized, due to the low cost of coal and continuing concerns about nuclear safety.

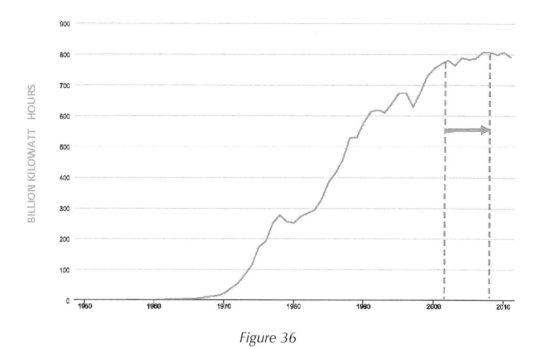

Figure 36

The Energy Policy Act of 2005 provided tax incentives for renewable energy, such as wind and solar power. In some parts of the United States, the cost of electricity from wind energy was less than the cost of electricity from nuclear energy.

The rapid growth of wind energy continued in the United States, as shown in *Figure 37*. The Energy Policy Act of 2005 provided for tax incentives for both wind and solar energy. However, the cost to generate electricity for power plants from wind energy was significantly cheaper than solar power in 2008. It would take another seven years before solar energy would begin to grow at a rate like wind energy.

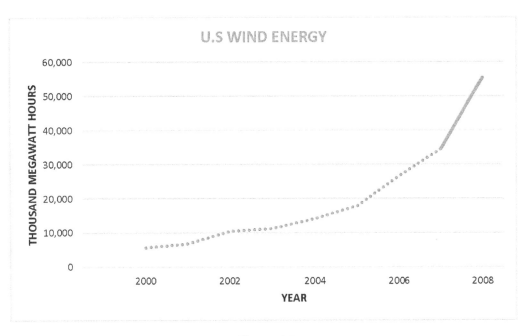

Figure 37

CHAPTER 16

Fresh Out of Miracles in Oz

Barbara and I had visited Australia several times on holiday while we were living in Indonesia and Malaysia. We always enjoyed the beautiful beaches, unique wildlife, and friendly people on our holidays. However, we didn't really appreciate the vastness of the country until we lived in the city of Perth in the state of Western Australia, as shown in *Figure 38*. It takes approximately five hours to fly from Perth to Sydney, Melbourne, or Brisbane on the east coast of the country.

Figure 38

Australia is an arid country that has a population of 23.8 million people. Western Australia is the largest state in the area but has a population of only 2.6 million people. The population of the three eastern states of Queensland, New South Wales, and Victoria is 18.7 million people. Many people in Western Australia believe their state is ignored by the federal government because their state's population is significantly smaller than the population of the eastern states. Over the next three years, I would come to understand why many people in Western Australia resented the federal government in Canberra.

In 2007, Prime Minister John Howard with the Liberal Party was soundly defeated by Kevin Rudd with the Labour Party. John Howard had been prime minister since 1996. Prime Minister Rudd won the 2007 election on a slogan of "new leadership" for the country.

In my opinion, the Australian political system is more focused on the polls in the news than solving the country's issues. As an example, the Australian political system doesn't allow experts in their field to be selected for ministry positions. In Australia, the prime minister selects members of parliament (MP) from his own party to the various ministry positions. A minister of energy may be an MP from the state of New South Wales and know absolutely nothing about the country's energy issues.

In the United States, the president usually selects his cabinet based on expertise in their field, such as energy, commerce, education, etc. The cabinet positions are closely gazetted by both Democrat and Republican senators. I saw, firsthand, Australian energy policies set in place by politicians with no knowledge or expertise in the country's energy issues.

In August 2008, Barbara, Ziggy the cat, and I arrived in Perth. The financial crisis in the United States started in 2007 due to failure in the subprime mortgage market. In 2008, the American economy had gone from bad to worse. However, the Western Australia economy was booming due to exports in iron ore, liquified natural gas, gold, alumina, and nickel to Asia—primarily China.[51]

The housing market in Perth was limited and expensive, but Barbara found a small apartment near the magnificent King's Park. Our apartment was in a great location, as I could either jog or take a fifteen-minute bus ride to the office.

Murphy Oil's president had authorized the formation of a small new ventures team in Jakarta, Indonesia. The new ventures team's focus was Southeast Asia and Oceania. The team was under tremendous pressure to find new opportunities, which resulted in the acquisition of two high-risk exploration leases in Australia and two even higher-risk exploration leases in Indonesia.

---

51   Economy of Western Australia - Wikipedia

Murphy Oil Australia's two offshore exploration leases were in the Browse Basin, as shown in *Figure 39.*[52]

Figure[53] *39*

Our office was in a modern, multistory office building in downtown Perth. The exploration team consisted of an experienced geophysicist with whom I had worked in Malaysia, an information technology specialist, and a part-time geological consultant. The job market in Perth was extremely tight, but I was fortunate to recruit another experienced geologist.

Murphy Oil acquired the first of the two leases in the Browse Basin by farm-in from a small Australian independent company. When I arrived in Perth, the partnership had agreed to drill one exploration well in the lease, AC/P36. The rig would arrive and spud the exploration prospect, Abalone Deep, in September 2008.

The exploration team's review of the prospect raised two significant concerns. The first concern was that the objective of the Abalone Deep exploration prospect

52   Geoscience Australia
53   Courtesy of GeoEdges Inc.

had no significant reservoir quality rock in the any of the wells within twenty miles of AC/P36. In fact, I couldn't find any fields in the Browse Basin or the Bonaparte Basin which produced hydrocarbons from the primary objective in the Abalone Deep exploration prospect.

The second concern was the economic viability of the Abalone Deep exploration prospect. The hydrocarbon type in the AC/P36 Block was almost certainly going to be natural gas. The reserve potential of the exploration prospect was modest and certainly wouldn't support building a pipeline to shore to sell gas on the local market.

My two concerns were quickly dismissed by Murphy Oil's president. We spudded the well on schedule and in November 2008 plugged and abandoned Abalone Deep as a dry hole. In December 2008, the world was engulfed in a financial crisis. Oil prices dropped from $103.94 per barrel in September 2008 to $41.44 per barrel in December 2008. I thought Murphy Oil might close the Australian office.

Organizationally, Murphy Oil's CEO stepped down but remained on the board of directors. Our president was then elected to the position of CEO. I suspected our new CEO wouldn't want to close any existing offices. We also had a new executive vice president arrive in Perth, and I now had another new boss.

The exploration team's review of the second lease, WA-423-P, was also concerning. The geological risks on the WA-423-P prospect were even greater than the risks on the Abalone Deep prospect, which had been a dry hole. In my opinion, the probability of finding commercial quantities of hydrocarbons in WA-423-P were almost nonexistent.

Murphy Oil's future in Australia didn't look very bright with our limited, high-risk prospect inventory. However, our company had a strong balance sheet, due to the oil and gas discoveries in Malaysia. The global financial crisis was causing many independent companies to sell assets. I thought the financial crisis might provide Murphy Oil an opportunity to acquire quality assets.

The first opportunity I pursued was the sale of the Montara Field by Cogee Resources. Cogee Resources were marketing several small oil discoveries in the Browse Basin. Murphy Oil had done similar small oil-field developments in Malaysia and in the Gulf of Mexico. I thought the Montara Field might be the perfect opportunity to leverage Murphy Oil's expertise.

I learned that Cogee Resources was financially distressed. Cogee Resources was attempting to sell the Montara Field in a sealed bid auction. I called the Cogee

Resources managing director and posed a hypothetical question, "If your company doesn't get a buyer for the Montara Field, would you consider selling these assets to Murphy Oil in return for a royalty on future production?" He responded, "No one thinks we will receive a single offer for the Montara Field. If I have no offers then yes, I would absolutely consider such a commercial deal."

Two days later, I learned that PTTEP, the national oil company of Thailand had paid approximately one hundred seventy million dollars for Cogee Resources, which consisted of the Montara Field.[54] I called the Cogee Resources managing director and congratulated him on the sale. I deduced from our conversation that PTTEP was the only bidder for Cogee Resources.

The second opportunity I pursued was an undeveloped gas field in the Bonaparte Basin. Santos, an Australian independent company, was selling select gas assets, and I thought this field would be commercial if the natural gas was sold to an existing LNG plant.

The Energy Business Group of the Mitsubishi Corporation was one of our partners in the exploration lease, WA-423-P. Mitsubishi was renowned as a leader in all aspects of LNG. I met with their representative and asked about the feasibility of monetizing the undeveloped Santos gas field. Their assessment was that it was possible, but only if we could acquire the field for a minimal price. We agreed that Murphy Oil would go to the Santos data room and complete a thorough reserve assessment of the undeveloped field.

Apparently, the concept of submitting a minimal bid on the undeveloped gas field was submitted to senior management at Mitsubishi. My boss in Perth and I discussed our proposal with our CEO, complete with a thorough reserve assessment and economic analysis. Our CEO quickly rejected our proposal without an explanation. Mitsubishi came back stating they were ready to submit a bid on the undeveloped gas field. The discussions with Mitsubishi can only be described as uncomfortable.

Mitsubishi then proposed flying a team to El Dorado, Arkansas, for a comprehensive presentation on LNG to Murphy Oil senior management. Our CEO agreed with Mitsubishi's proposal. I knew that our CEO and select members of the Murphy Oil board of directors were impressed with the presentation. However, after the meeting I was told that Murphy Oil was no longer interested in LNG in Australia.

The CEO then called to tell me to pursue oil opportunities in the Bass Basin and Gippsland Basin in southeastern Australia. Our CEO had been involved in new ventures in Australia prior to joining Murphy Oil, and he thought he could

---

54   Offshore Technology – Montara Oil Field, Timor Sea

point me in the right direction. I agreed with our CEO's assessment that the Bass Basin and Gippsland Basin had excellent oil potential. Unfortunately, every area our CEO had identified was held by producing oil fields operated by ExxonMobil. We both knew that ExxonMobil was not going to sell any quality producing asset in Australia or anywhere else in the world. I also reminded our CEO that over 80 percent of Australia's reserves are natural gas, and quality oil opportunities are few and far between. My assessment of Australia's oil potential was not well received by Murphy Oil's CEO.

Over the next year, the exploration team evaluated numerous asset acquisition and exploration opportunities. After many dead ends, we developed an exploration oil play in the offshore Perth Basin. The exploration play was an extension of the proven oil and gas fields onshore. However, the limited available data indicated the potential for larger structures, which could mean commercial oil or gas fields.

Unlike the United States, Australia doesn't have natural gas pipelines crisscrossing the country. In the eastern states of Australia, the primary energy source for power plants was coal. Coal bed methane provides fuel gas in the homes. In Western Australia, the primary energy source for power plants is natural gas.

Although the northwest shelf of Australia has multiple giant natural gas fields, the gas from these fields was committed to massive LNG plants for export. As a result, natural gas prices in Western Australia ranged from $8 to $9 per standard cubic feet (scf). In Eastern Australia, coal-bed-methane gas sold for $1.25 to $1.50 per scf.

The hydrocarbon potential in the offshore Perth Basin could be oil and/or natural gas. The economics on any reasonable-sized discovery would be robust, whether it was oil or natural gas. Our CEO initially pushed back about pursuing exploration in the offshore Perth Basin. However, he quickly changed his mind once he understood the prices for natural gas in Western Australia.

I flew to Canberra to meet with Geoscience Australia to discuss the offshore lease sale process, timing of the next lease sale, and possibility of nominating acreage in the offshore Perth Basin. I learned that the lease sale process was straightforward and transparent. Oil and gas companies nominate acreage, which is then reviewed by Geoscience Australia. After the review, the acreage is then posted for bids for the upcoming lease sale.

There was already one small oil field in the offshore Perth Basin. The area Murphy Oil wanted to nominate for the upcoming lease sale had been previously leased in the 1970s. There were no environmental issues with the area we wanted to nominate.

No one at Geoscience Australia anticipated any problem with a quick review and posting of Murphy Oil's area of interest for the next lease sale.

Murphy Oil conducted an independent study to ensure there were no environmental issues with our area of interest. I included the independent environmental study with the lease nomination documents. I also included a letter from the premier of Western Australia, Colin Barnett of the Liberal Party, supporting Murphy Oil's lease nomination and the potential positive impact a gas discovery would have on energy security and the economy.

I received a very disturbing call from Geoscience Australia two weeks before the federal government was going to post acreage for the upcoming lease sale. I was told Murphy Oil's area of interest wasn't going to be nominated for the upcoming lease sale. I couldn't get a reason from Geoscience Australia why the acreage wouldn't be posted. I met with one of the MPs for Western Australia, who recommended I go to Canberra and meet with Geoscience Australia, the Labour government department head that oversaw Geoscience Australia, and the Liberal shadow cabinet energy minister to discuss the lease-posting process. I received a similar opinion from the office of the premier of Western Australia. I flew to Canberra the next day to meet with the various government officials.

In my first meeting, I was told by the "acting head" of Geoscience Australia that no acreage for the offshore Perth Basin would be posted until Geoscience Australia had done a comprehensive study to confirm the potential for hydrocarbons. I pointed out that there was already one producing oil field in the offshore Perth Basin and numerous oil and gas fields on the onshore region of the Perth Basin. I was told by the acting head of Geoscience Australia, "Sorry, my decision has been made."

The technical managers for Geoscience Australia came up to me after the meeting and apologized for such a daft response from the acting head. They also said it would be an absolute waste of their limited resources to do such a study for the offshore Perth Basin. However, they would have to comply with the acting head's directive. They also confirmed there were no environmental issues with Murphy Oil's area of interest.

I then met with the shadow cabinet minister of energy from the Liberal Party. He was very well informed on the energy issues in Western Australia, as well as the entire country. He said he saw no logical reason not to nominate Murphy Oil's acreage at the upcoming offshore lease sale. He also commented that politics can sometimes get in the way of common sense. His comments would prove to be prophetic.

Finally, I met with the newly appointed Labour Party department head, who oversaw Geoscience Australia and the licensing round process. In preparation for the meeting, I learned the department head's only energy-related experience was as a lobbyist for Chevron. The lobbyist's role was to gain government approval for Chevron's two massive LNG projects in Western Australia. In this meeting, it quickly became apparent the department head knew absolutely nothing about even basic energy issues for the country. It was clear the department head had absolutely no interest in doing anything to address any energy issue in Western Australia.

I left these meetings convinced the offshore licensing round had become a political battle between the Labour Party and the Liberal Party. It would take Geoscience Australia and the Labour government another two years before any offshore Perth Basin acreage was posted for a lease sale. The people of Western Australia continued paying exorbitant prices for natural gas.

As I was flying back to Perth, I reflected on my experiences with Australian politics. I think Australian politics can be described as all-out warfare. The political warfare can be with the opposition party or even within their own party. In June 2010, Labor MP Julia Gillard ousted Kevin Rudd as head of the Labour Party and became prime minister of Australia. In June 2013, then Labor MP Kevin Rudd ousted Julia Gillard as head of the Labor Party and became prime minister.

After three years, I saw no way to meet our CEO's objectives in Australia. I was completely frustrated with the Australian government and the licensing round process. I elected to take retirement from Murphy Oil, effective February 2001.

In January 2011, as Barbara and I were preparing to move from Perth, Australia, to Portland, Oregon, the average price for oil in the United States was $87.04 per barrel. From 2008 to 2011, the price of oil had decreased by $4.44 per barrel. In 2011, the United States was producing 5.64 Million Barrels of Oil per Day (MBOPD), while consuming 18.88 MMBOPD a day.

From 2008 to 2011, the amount of oil produced in the United States increased by over 645,000 Barrels of Oil per Day (BOPD). From 2008 to 2011, the consumption of oil in the United States decreased by more than 600,000 BOPD, as shown by Figure 40.[55]

---

55 U.S. Energy Information Administration – Total Energy Review

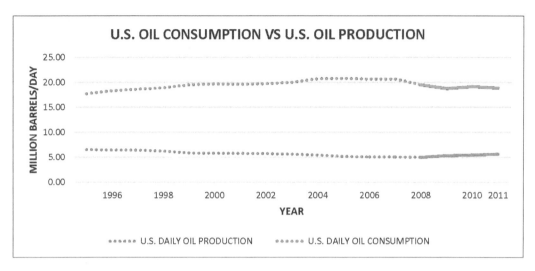

Figure 40

The increase in oil production was a result of small, dynamic oil companies applying shale oil technology developed by the service companies like Schlumberger and Halliburton in the United States. Horizontal drilling and hydraulic fracturing ("fracking") represented a technology revolution in the hydrocarbon industry. This revolution would result in the United States once again becoming a leading hydrocarbon producer in the world.

In 2011, the average price for gasoline in the United States was $3.53 per gallon, which is equivalent to $3.75 per gallon, when the price is adjusted to inflation (March 2015). From 2008 to 2011, the price of gasoline in the United States had increased only $0.26 per gallon. The decrease in oil consumption from 2008 to 2011 was due to continued high oil and gasoline prices. History has proven there is a direct correlation between energy conservation and energy prices.

In the People's Republic of China (PRC), the consumption of oil was continuing to rapidly increase, outpacing the domestic supply of oil as shown in *Figure 41*.

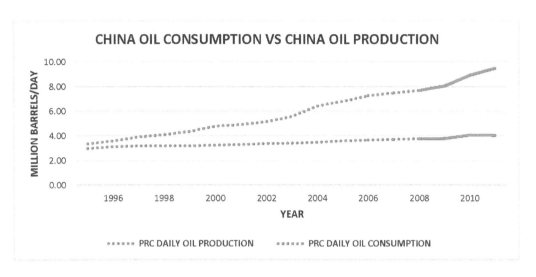

Figure 41

The unconstrained industrialization of the People's Republic of China (PRC) which started in the 1980s, created toxic air quality in all their cities. In 2000s, PRC started shifting from coal to natural gas for the fuel for their power plants. Natural gas emits significantly less greenhouse gases than coal.

The PRC purchases natural gas from pipelines originating from Uzbekistan, Kazakhstan, and Myanmar, and LNG from Australia, Indonesia, Malaysia, and Qatar. In 2011, the PRC was importing approximately twenty billion cubic feet of gas a day.[56]

The PRC's increase in natural gas and LNG imports resulted in a decrease in coal imports. The decrease in the PRC's coal imports impacted coal production in the United States. From 2008 to 2011, the annual coal production in the United States dropped by approximately one million short tons, as shown in *Figure 42*.[57]

56 Natural Gas in China a Regional Analysis, The Oxford Institute for Energy Studies, November 2015
57 U.S. Energy Information Administration – Total Energy Review

FRESH OUT OF MIRACLES IN OZ | 183

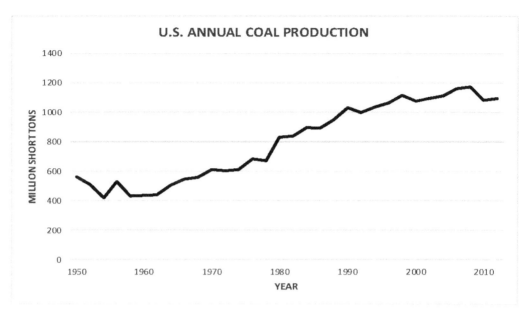

Figure 42

From 2008 to 2011, the rapid growth of wind energy continued in the United States, as shown in *Figure 43*. Wind turbine technology continued to improve, making wind energy more reliable and cheaper than nuclear energy. The tax incentives from the Energy Policy Act of 2005 made wind energy cost competitive with oil and natural gas. However, it would take another four years before solar energy would begin a growth period, like wind energy in 2005.

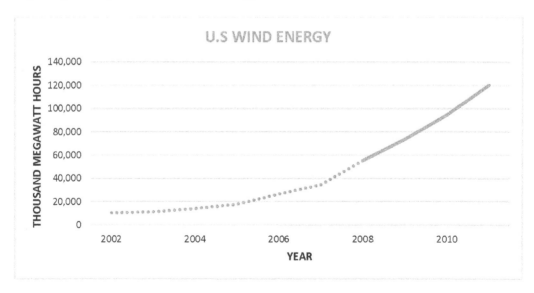

Figure 43

CHAPTER 17

Sunset in Arabia

Barbara and I had decided to retire in Portland, based on the climate, cultural diversity, and natural beauty of the area. Once we arrived in Portland, we immediately set about looking for a home. In April 2011, we moved into a beautiful home in the Portland Hills, overlooking the Willamette River and the Cascade Mountains.

Although I received several calls about returning to work, I thought it might be time for me to pursue other interests, like additional studies at university. After a few months, I was contacted by a former colleague who told me of an exploration manager position in Muscat, Oman. He thought I would be perfect for the position. I certainly enjoyed my brief time in Ankara, Turkey, and I thought this might be an opportunity to learn more of the different cultures in the Middle East. I asked for more information on the company and the company's operations in Oman.

Within a few days, I received information on the company, Consolidated Contractors Energy Development (CCED). The company was a privately owned oil and gas production company that was established in 2007. CCED was a subsidiary of Consolidated Contractors Company (CCC), a well-known international construction company.

After I received the information on CCED, I started researching the Sultanate of Oman. The country is located on the southeastern area of the Arabian Peninsula, as shown in *Figure 44*. Sultan Qaboos bin Said al Said has been the leader of the country since 1970. The sultan appoints a cabinet, *Diwans*, to assist him in the governance of the country.[58] In the early 1990s, the sultan instituted an elected advisory council, *Majlis ash-Shura*. I interpreted the *Majlis ash-Shura* as the first step in the introduction of democracy to the Sultanate of Oman.

---

58 Oman Sultanate Politics - www.omansultanater.com/politics.htm

Figure 44

Sultan Qaboos was highly educated, having graduated from Sandhurst in the United Kingdom. He modernized the country, building schools, hospitals, roads, and airports. Although officially neutral, Oman had normal relations with Iran and was also a close ally of the United Kingdom and the United States. The more I read, the more Sultan Qaboos reminded me of the dynamic former prime minister of Malaysia, Dr. Mahathir bin Mohamad.

I received very little technical information from my former colleague on the two licenses CCED operated. However, CCED had two oil fields, which were producing approximately 2,000 barrels of oil per day. After discussing the position with Barbara, I submitted my resume for the position of exploration manager.

One week later, I was told CCED's CEO wanted to interview me by telephone. In the telephone interview, the CEO's questions me questions about my technical background and leadership experiences. The CEO told me that from 1972 to 2008, eight different companies explored the area currently held by CCED. None of the previous companies had any commercial success drilling in the area. In 2009, CCED's first two exploration wells were commercial oil discoveries. The CEO spent the next thirty minutes answering my questions on the two producing fields, partners, government, and living in Oman.

At one point in our discussion, the CEO said, "I'm not sure who is interviewing whom." I replied that I didn't want to take a position unless I believed I could make a positive impact. The CEO then said, "Would you have any problem agreeing to a three-month probationary period?" I told him I liked his idea, as it was a probationary period for both CCED and me. The CEO then laughed and said he would send me the job offer the following week.

On July 1, 2011, I flew from Portland to Muscat, Oman. After spending over twenty-four hours in the air, I arrived at Muscat International Airport, tired but excited about the new opportunity facing me. The next morning, a CCED driver picked me up at the hotel and took me to the office. CCED's office was only one floor in a modern, multistory office building. The office building was just a few blocks from Sultanate of Oman's Ministry of Oil and Gas (MOG) and the United States Embassy.

Once I arrived at the office, I met the CEO, my new boss. I would characterize my new boss as personable, driven, and technically astute. He provided me with a detailed assessment of the two producing fields, Farha South Field in Block 3 and Saiwan East Field in Block 4, as shown on *Figure 45*. He concluded by saying, "Your new team has some unique personalities that you may find challenging."

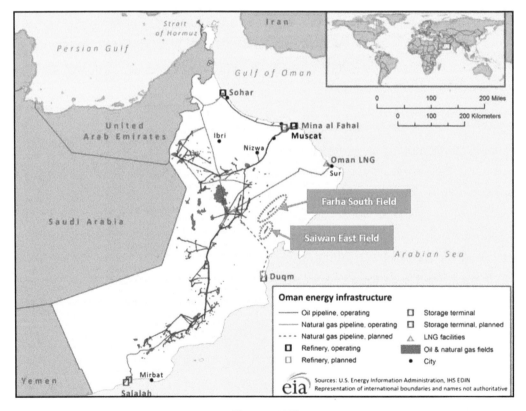

Figure 45[59]

---

59   Courtesy of U.S. Energy Information Administration (Oct 2008)

CCED had approximately thirty-five people in the entire office. The exploration department consisted of two expatriate geologists and two recent graduates from Sultan Qaboos University (SQU). The two recent graduates were bright but had been doing primary geotechnical support work for the two expatriate geologists. It would take at least one year of intensive training and mentoring before the two recent graduates would be able to do independent geological interpretation and analysis.

One of the expatriates was working as a geophysicist, because he knew how to operate a workstation. He was responsible for the geophysical interpretation of the large 3-D seismic surveys over two producing fields. It quickly became apparent to me, that he had only a rudimentary understanding of geophysics. I knew I would have to closely check all his work prior to submitting any drilling recommendation.

The second expatriate was an experienced and talented well-site geologist. I knew he would be an invaluable resource during our drilling operations. However, he was going through a very difficult divorce, which impacted his temperament and his productivity.

My predecessor had been fired by the CEO. As I went through the office files, I could find absolutely no evidence my predecessor had done any work over the previous year. No one in the exploration team was sure what he had done over the previous year either.

Operationally, CCED had two drilling rigs under contract for the next two years. Either rig could drill a typical well in three to four weeks. When I arrived, CCED had a prospect inventory of zero and an expatriate geologist with a bucket full of excuses as to why there were no prospects. More than once, I thought I should just get on a plane and go home.

Fortunately, we had two very supportive partners, Tethys Oil and Mitsui Exploration and Production. The partners only request was to provide more lead time to review any well proposal. I agreed with their request, but asked them to give me time, as our exploration team was resource challenged. The partners nodded, knowingly.

Over the next three weeks, I spent at least six hours a day with the geophysicist developing four independent prospects, which would extend the Farha South Field another ten miles north. I then wrote each of the well proposals, which contained a probabilistic reserve assessment and a probabilistic chance of commercial success. Our CEO was very happy with the four prospects and the changes to the well proposal. Our partners were ecstatic with an actual prospect inventory and the format of the new well proposal. I was learning once again how to function with only four hours of sleep a night.

As I was starting to feel we were making progress, I received a letter from Petroleum Development Oman (PDO) inquiring on the status of a mega-data trade, initiated by my predecessor. PDO is owned by the Government of Oman (60%), Royal Dutch Shell (34%), Total (4%), and Partex (2%). Royal Dutch Shell has had a very strong influence in the organizational structure and culture of PDO, similar to their influence on PETRONAS in Malaysia.

My first impression was that a data trade of this size would shut down our small exploration team. It takes a significant amount of time to prepare digital and paper copies of just one log, let alone fifty wells with core data, oil samples, seismic data, etc. After further review, I thought none of the PDO wells had any immediate value to CCED's current technical evaluations. I now knew that my predecessor had done some work, though none of the work would help CCED!

PDO had copied the Ministry of Oil and Gas (MOG) in their letter on the data trade. The MOG is the government body in the Sultanate of Oman responsible for developing and implementing the government policies for exploiting the oil and gas resources in the country. I knew that if I didn't act quickly, I would have the MOG dictating the terms of the proposed data trade.

I discussed the data trade with the CEO who asked our Commercial Manager to meet with the MOG and ask for time for CCED and PDO to work out the details of the data trade. I sent PDO an e-mail requesting a formal face-to-face meeting to progress the proposed data trade. I quickly received a response from PDO proposing a date for a meeting.

The meeting was to be at PDO's office due to their exploration supervisor's "busy schedule." One week before the meeting, I received a request from PDO to fax my passport and work permit to their security team, who in turn would provide me a PDO security pass. In my reply, I declined to fax them these documents, due to my concern over security fraud. I heard nothing further from PDO's security team.

On the day of the meeting, I drove to the PDO compound and was stopped by their security team at the gate. I showed them the e-mail from the exploration supervisor setting up the date and time of the meeting. The PDO security team told me I couldn't enter because I didn't have a PDO security pass. I said, "Fine, now you call the exploration supervisor and tell her I won't make the meeting. Please explain that you are the reason I am not able to make this meeting." This response caused the security team to huddle and after a few minutes they told me to drive into the compound.

I arrived at the designated building, went inside, and asked to see the exploration supervisor. Again, I was told that I didn't have a PDO security pass. Once again, I said "Fine, I'll leave, but please call the exploration supervisor and tell her why I wasn't able to make this important meeting." The receptionist immediately called the exploration supervisor, who came running down to meet me. The exploration supervisor couldn't believe I had gotten through PDO's impenetrable security system. It was hard for me not to laugh.

The exploration supervisor was a secondee from Royal Dutch Shell. With her was the PDO chief geologist, who was also a secondee from Royal Dutch Shell. After a brief introduction, the PDO duo started asking me if I was a CCED employee or consultant. They wanted to know if I had found a home and how long I planned to stay in Oman. Being a mischievous person, I told them I was a permanent employee for CCED, and I was so impressed with Oman I was thinking of applying for citizenship. The shocked look on their faces told me they didn't realize I was kidding about citizenship. Apparently, the Royal Dutch Shell exploration manager I knew in Malaysia had been transferred to Oman. Royal Dutch Shell's missteps in Malaysia were well known inside their company and apparently so was I.

In my PDO meeting, we finally moved to the topic of the data trade. I told them CCED was very pleased to do a data trade, but I only wanted six PDO wells and would reciprocate with a like number of CCED wells. I also explained what data CCED could provide. When I was asked about additional data I replied, "I can't trade what I don't have." I sent PDO and copied the MOG a summary of our meeting minutes. Fortunately, I was able to diffuse a potential shutdown of our exploration department from a poorly conceived data trade.

The Farha South Field consisted of seven small, tilted fault blocks, which produced oil from a sandstone reservoir. I saw dozens of undrilled similar fault blocks on the 3-D survey, which for some reason had been downgraded by the geophysicist. Our new prospect inventory would soon test four of the undrilled structures in the Farha South Field.

All four of the prospects in the Farha South Field were commercial oil discoveries. Each well was producing oil in less than three weeks after the well was drilled. Our daily oil production quickly climbed to over 5,000 barrels of oil per day (BOPD).

With each new oil discovery, we gained two to eight new development locations, depending on the size of the fault block. The development wells consisted of water injector wells and additional oil producers to maximize the recovery of the oil from each fault block. I built a simple spreadsheet to track the status of each future well location. The spreadsheet proved to be an invaluable communication tool with the drilling department and the facilities department, who were responsible for building the drilling site and then installing the oil flow lines to our pipeline system.

As our daily oil production increased, our office started to expand and CCED took over new floors in our office building. I was given approval to hire new exploration staff. Our exploration department soon looked like the United Nations with representatives from Egypt, Malaysia, Oman, Palestine, Syria, United Kingdom, United States, and Yemen. More importantly, I was able to recruit recent national graduates from SQU and provide them with a quality training and mentoring program. The two graduates that were at CCED when I first arrived were now doing quality, independent technical work.

I also purchased another discovery bell, which we rang every time we had an exploration discovery. The exploration team that developed the prospect were recognized by their peers for their hard work. The exploration team also gave the other departments an overview of the discovery. After our ceremony, I would summarize what the discovery would mean to the employees of CCED and the financial benefit to the people of Oman. The discovery bell seemed to energize the entire exploration team.

I quickly found out that the Omani's enjoy a good cup of coffee and sweets. As a result, we always had cake to celebrate every exploration discovery. I suspect the enjoyment of cakes and sweets is why there is a dentist on almost every street corner in the city of Muscat.

At the end of 2011, I discovered that CCED had committed to drill an exploration well outside of the large 3-D surveys over the Farha South Field and the Saiwan East Field. This meant that CCED would have to use the vintage 2-D seismic data that that been acquired by the previous operators to develop an exploration prospect in either Block 3 or Block 4. I was concerned about the accuracy of the location of each seismic line. I had no way to authenticate the location of any of the legacy seismic lines. We finally developed one prospect, less than one mile southwest of the Farha South Field 3-D survey.

In the Farha South Field, we were able to predict the top of the objective within fifty feet because of the modern 3-D seismic data. However, the exploration well drilled on the vintage 2-D seismic data provided me a very unpleasant surprise. Our reservoir objective came in five hundred feet low to the predrill well prognosis. We sidetracked the exploration well, anticipating we would drill to a structurally higher location in the fault block. To my horror, we came in over seven hundred feet low to the revised predrill well prognosis. This told me there was a significant error in the positioning of the vintage 2-D seismic data.

I asked our facilities department to locate the drilling location of several of the wells drilled by the previous operators. They found that none of the wells drilled by the previous operators were in the location reported in the company's final well report. The errors in the positioning of the legacy wells ranged from one hundred to one thousand feet. I expected the 2-D seismic positioning errors to be of the same magnitude as the well locations. Positioning errors must have been a major contributor to the numerous dry holes drilled by the previous eight companies.

At the end of 2011, I completed the probationary period with CCED and was offered a two-year employment contract. Over the previous months, the CEO and I had become friends. I had also gotten to know one of the owners of the company, who was a very kind and gracious man. I also felt that I was making a difference in the company and enjoyed my work at CCED.

Barbara and I discussed the situation and I agreed to sign the employment contract. Barbara joined me in Muscat, Oman, in 2012. We were able to find a modern, furnished apartment just a few miles from CCED's office. We have many fond memories of Oman. We were always impressed with the warm and courteous people of the country.

In 2012, we planned a major 3-D seismic program in Block 3 and Block 4. The MOG requires all major projects be tendered. The MOG wants to ensure that only a qualified company with the lowest price is awarded the program. CCED tendered the 3-D seismic program and awarded the contract to BGP Inc. The company is owned by China National Petroleum Corporation (CNPC) and is recognized as a leader in onshore seismic acquisition around the world.

BGP Inc. has an excellent quality control program, starting with a daily check of the tens of thousands of geophones that are used every day, as shown in *Figure 46*. The company used the latest equipment, such as the vibrators shown in *Figure 47*. The data BGP Inc. acquired was of the highest quality. More importantly, the company achieved an exemplary safety record while operating in scorching desert conditions.

Figure 46[60]

Figure[61] *47*

Our exploration program continued to be very successful. In the Farha South Field, we discovered oil in ten new fault blocks. We also had a new exploration discovery just two miles east of the Saiwan East Field. CCED's oil production was now

60 Courtesy Mokhles Ahmad
61 Courtesy Mokhles Ahmad

more than 15,000 BOPD.

The new 3-D surveys helped the exploration team significantly increase our exploration prospect inventory. Global exploration commercial success rate for major and independent companies range from 25% to 35%. CCED's our exploration commercial success rate was over 70%. The exceptionally high commercial success rate told me Block 3 and Block 4 had significantly more undiscovered oil potential.

In 2013, we started our exploration drilling program in the western region of Block 4, south of the Farha South Field and west of the Saiwan East Field. The name of the first exploration well in this area was *Shahd*, which means "honey" in English.

The Shahd Prospect turned out to be very sweet for CCED. The exploration well had multiple oil pay zones and opened new exploration plays in Block 3 and Block 4. Six more new oil fields were found within a ten-mile radius of the Shahd Field. CCED's daily oil production now exceeded 25,000 BOPD, a tenfold increase from when I arrived in July 2011.

As 2013 was drawing to a close, I was faced with personal issues. Barbara's mother was not well, and she wanted to return home to be closer to her mother. With regret, I submitted my resignation to CCED. The CEO asked me to work as a consultant, which I agreed to do until 2016. One of my fondest memories is of our exploration team, shown in *Figure 48*.

Figure 48[62]

---

62   Courtesy Mokhles Ahmad

In November 2013, as I was preparing to move from Muscat, Oman, back to Portland, Oregon, the average price for oil in the United States was $91.17 per barrel. From 2011 to 2013, the price of oil had increased by $4.13 per barrel. In 2013, the United States was producing 7.44 Million Barrels of Oil per Day (MMBOPD), while consuming 18.96 MMBOPD a day.

From 2011 to 2013, the amount of oil produced in the United States increased by over 1,795,000 Barrels of Oil per Day (BOPD). From 2011 to 2013, the consumption of oil in the United States increased by less than 80,000 BOPD, as shown by *Figure 49*.[63]

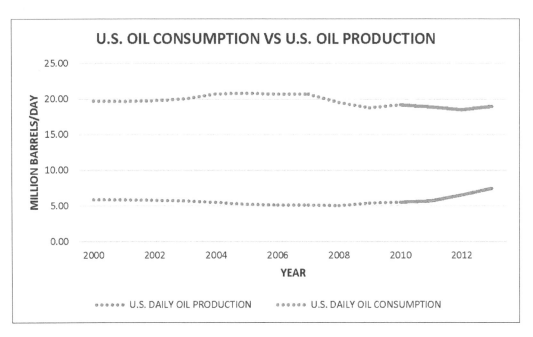

Figure 49

The dramatic increase in oil production was a result of small, dynamic companies rapidly applying shale oil technology from the service companies in the United States. Horizontal drilling and hydraulic fracturing ("fracking") were revolutionizing America's hydrocarbon industry.

The United States is the only country that has significantly benefited from shale oil technology. I believe the shale oil boom in the United States is due to several factors, including the significant number of petroliferous basins, business conditions,

63   U.S. Energy Information Administration – Total Energy Review

and the hundreds of small, dynamic companies operating in this country.

In 2013, the average price for gasoline in the United States was $3.53 per gallon, which is equivalent to $3.62 per gallon, when the price is adjusted to inflation (March 2015). From 2011 to 2013, the price of gasoline in the United States didn't change. There was a very minor increase in oil consumption from 2011 to 2013, which is probably due to the beginning of an economic recovery in the United States.

In the People's Republic of China (PRC), the consumption of oil was still rapidly increasing, as shown in *Figure 50*. The domestic supply of oil in the PRC had effectively plateaued.

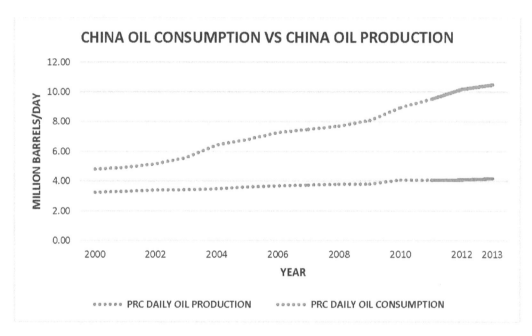

Figure 50

The PRC continued to increase natural gas imports while decreasing coal imports to improve the air quality in their major cities. In 2013, Europe's coal imports from the United States also began to decline. Europe and the PRC were actively seeking alternatives to coal. The reduction in coal exports to the PRC and Europe negatively impacted U.S. coal production, as shown in *Figure 51*.[64] Coal production began to steadily decline, ending the coal production boom in the United States.

---

64   U.S. Energy Information Administration – Total Energy Review

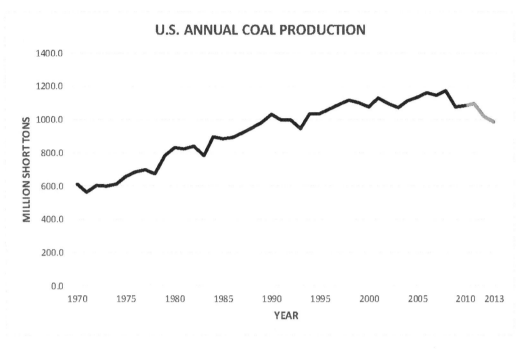

U.S. ANNUAL COAL PRODUCTION

Figure 51

The rapid growth in wind energy also continued in the United States from 2011 to 2013, as shown in *Figure 52*. Wind turbine technology continued to improve, making wind energy more cost competitive with oil and natural gas.

In my opinion, 2012 was the start of the solar energy boom in the United States, as shown in *Figure 53*. The dramatic increase in solar energy was primarily driven by the installation of solar panels on homes and businesses. In 2011, the PRC started mass producing solar panels, driving down the global prices. The decline in solar panel prices, coupled with tax incentives from the Energy Policy Act of 2005, were the primary causes for the rapid growth in solar energy in the United States.

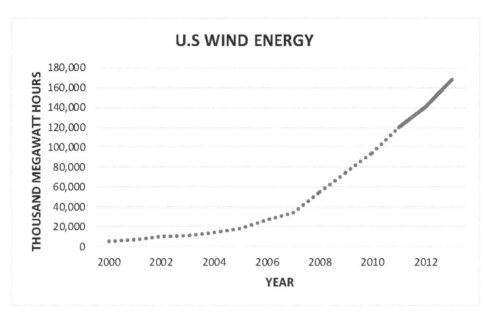

Figure 52

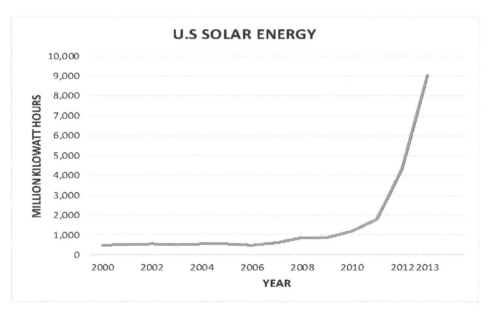

Figure 53

CHAPTER 18

Our Energy Conundrum

Fossil fuels (coal, oil and natural gas) were the life blood of the Industrial Revolution, which in turn elevated mankind's quality of life. Mankind's elevated quality of life also created an insatiable demand for sustainable, reliable, cheap energy. For over two hundred years, fossil fuels have met the world's energy demands. However, the unabated consumption of fossil fuels has created significant air quality issues in many regions around the world. A World Bank study estimated 3.7 million people died in 2012 from chronic or acute effects of atmospheric pollutants.[65] Most of the scientific community also support the theory that greenhouse gases are negatively impacting the earth's climate, causing global warming.

How will the world respond to the ever-increasing air pollution and the potential cataclysmic impact of climate change? Will the world's consumption of fossil fuel change in the next twenty to fifty years? In 2013, fossil fuels accounted for 81% of the fuel used in the world, as shown in *Figure 54*.[66] Wind and solar energy account for less than one percent of the fuel sources in the world.

---

65 "Enhancing the World Bank's Approach to Air Quality Management," World Bank, February 2015.
66 International Energy Agency 2015

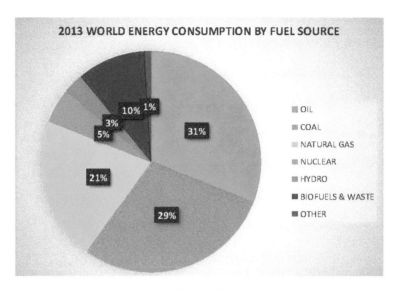

Figure 54

Oil is primarily used as a fuel for transportation (land, sea and air). Oil is also used to produce a wide range of products such as plastics, synthetic fabrics, cosmetics, and medicines. Oil is an insignificant fuel source for electricity generation in the world, as shown in *Figure 55*.

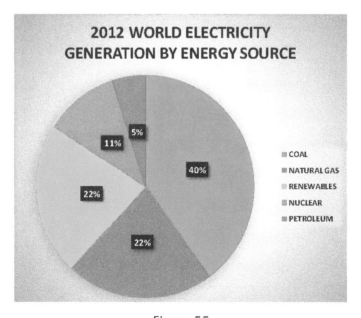

Figure 55

COAL

Forty percent of the electricity generated in the world today is from coal. Coal's dominance is due to its low cost relative to other fuels. However, coal generates the greatest amount of greenhouse gases of any fossil fuel—more than twice the amount generated by natural gas.[67]

The United States has the largest coal reserves in the world, with more than 25% of the proven global reserves.[68] The People's Republic of China (PRC) has the third largest coal reserves, with approximately 13% of the proven global reserves. However, the PRC is the largest producer and consumer of coal in the world, as shown in *Figure 56*.[69]

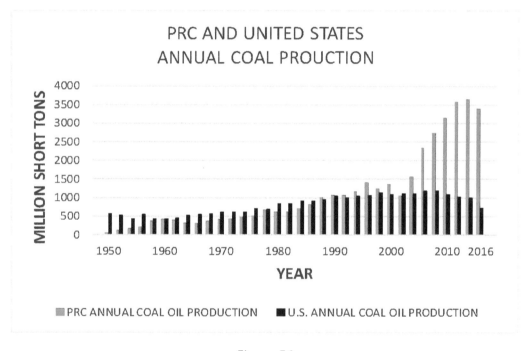

Figure 56

A report by the World Bank in cooperation with the PRC government found that approximately 750,000 people die prematurely in China each year from air pollution. The toxic air quality in the PRC has resulted in their government shifting from coal to natural gas in many of their power plants. The PRC is increasing the construction of nuclear power plants to replace inefficient coal power plants.

---

67 "Lifecycle Greenhouse Gas Emissions Estimates for Electricity Generators" by Benjamin K. Sovacool
68 "Countries with the Biggest Coal Reserves", miningtechnology.com November 21, 2013
69 National Bureau of Statistics of the People's Republic of China

Coal consumption in the PRC has decreased over the last three years.

Coal consumption is also decreasing in the United States and Europe. However, countries in need of low cost energy, such as India, are forecast to increase coal consumption over the next ten years.[70] Global coal consumption is forecast to decrease over the next ten years. However, the world has yet to demonstrate the discipline to walk away from cheap energy, even when it is in our own best interest.

OIL

Oil is the most used fuel source in the world. However, only five percent of the electricity generated in the world today is from oil. Oil generates approximately twenty percent less greenhouse gases than coal. Venezuela has the largest proven oil reserves in the world, with over 300 billion barrels, followed by Saudi Arabia with 269 billion barrels of oil, and Canada with 171 billion barrels of oil.[71] The United States and the PRC are the largest consumers of oil, at rates of 19.6 and 11.3 million barrels of oil per day respectively.[72]

Oil is primarily used as a fuel for transportation (automobiles, trucks, ships, and airplanes). Fuel economy in automobiles, trucks, and airplanes has improved. The average fuel economy for new vehicles purchased in 2013 was almost five miles per gallon better than vehicles purchased in 2007.[73] The Boeing 787 uses 40% less fuel than airliners used in the 1970s.[74] However, global consumption of oil has continued to increase, as shown in *Figure 57*.

70   "International Energy Outlook 2016," U.S. Energy Information Administration – May 11, 2016
71   "Top Ten Countries with World's Largest Oil Reserves," Energy Business Review – April 21, 2017
72   "World's Top Oil Consumers," by Irina Slav, OILPRICE.com – June 6, 2017
73   "Two Reasons Why Fuel Economy Is at An All-Time High," by M. Maynard, Forbes – September 10, 2013
74   "Changes in the Air," The Economist – September 2011

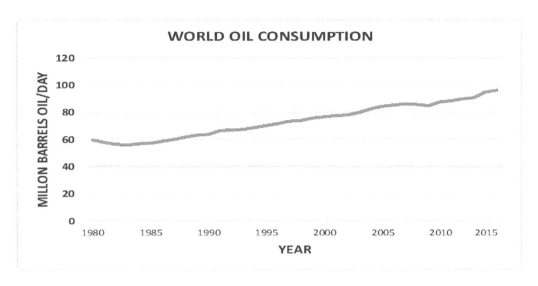

Figure 57

If the fuel efficiency for automobiles, trucks, and airplanes has improved, why has the world's consumption of oil increased by more than thirty-five million barrels a day since 1980? The answer is the rapidly growing number of vehicles, especially in developing countries like PRC and India.

A Ward's Auto report estimated there are now more than 1.2 billion vehicles in the world. The PRC has the most vehicles of any country in the world with 279 million vehicles, followed by the United States with 264 million vehicles. The number of vehicles in the world is forecast to double by 2040.[75] India's demand for diesel vehicles and China's demand for gasoline vehicles will continue to increase the demand for the supply of oil.

The demand for oil will continue to increase if oil is the primary fuel for automobiles, trucks, and airplanes. Electric cars offer a glimmer of hope, if the power plants use renewable energy. An electric car charged by a coal-fired power station may have a greater carbon footprint than a fuel-efficient gasoline powered automobile. Mass transit is another solution, but it requires reeducating people to break their addiction to gasoline-powered vehicles.

---

75   World Economic Forum by M. Smith Finance Writer – April 22, 2016

NATURAL GAS

Twenty-two percent of the electricity generated in the world today is from natural gas. Natural gas has less than 50% of the greenhouse gases of coal and less than 40% of the greenhouse gases of oil.[76] The Russian Federation has the largest proven natural gas reserves in the world with 1,688 Trillion cubic feet (Tcf), followed by Iran with 1,200 Tcf, Qatar with 866 Tcf, and the United States with 368 Tcf.[77] The United States is the largest consumer of natural gas in the world at a rate of 27.4 Tcf per year.[78]

Oil can be easily transported by truck, railroad, or pipeline from the field to the refinery or the power plant. Natural gas can is only transported by pipeline or as liquefied natural gas (LNG) to the power plant. North America and Europe have extensive pipeline systems in place to transport natural gas from fields in the region or from LNG terminals. The PRC is rapidly developing pipeline systems to import natural gas from Myanmar, Uzbekistan, and Turkmenistan. Japan, South Korea, and PRC are the three largest importers of LNG in the world.

In thirty-five years, the consumption of natural gas has increased by more than 230% , as shown in *Figure 58*.[79] Natural gas consumption in the world is forecast to continue to increase over the next forty years due to low price, abundant supply, and low carbon footprint for a fossil fuel.

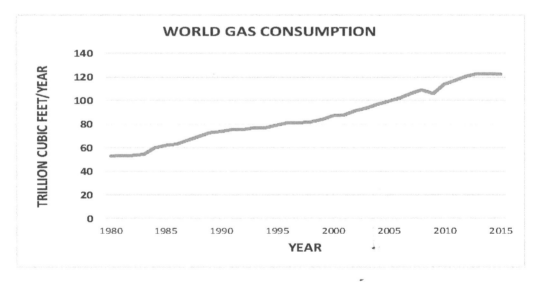

Figure 58

76 "Lifecycle Greenhouse Gas Emissions Estimates for Electricity Generators," by Benjamin K. Sovacool
77 "The World Fact Book," Central Intelligence Agency – January 1, 2016
78 U.S. Energy Information Administration
79 U.S. Energy Information Administration

Global warming is a theory, supported by the majority in the scientific community. However, many challenge the validity of global warming, including the current president of the United States, Donald Trump. If you are a global warming skeptic, why would you support the development of renewable energy, which is costlier than fossil fuel?

The world is currently awash in relatively cheap fossil fuel (coal, oil, and natural gas). However, the world has already consumed almost all of the highest quality coal. The world is consuming more than a thousand barrels of oil a second.[80] The world will eventually run out of sustainable supplies of fossil fuel.

When exactly will the world run out of fossil fuel? British Petroleum Statistical Review of World Energy 2016 estimates the world has 115 years of coal production, 100 years of natural gas production and less than 50 years of oil production. British Petroleum's forecast emphasizes fossil fuel forecasts will continue to vary with time due to fossil fuel consumption rates, global economic growth, etc. Although we don't know exactly when we will run out of fossil fuel, we do know we will eventually run out and probably in the not too distant future. Shouldn't America support the development of sustainable, reliable, and cost effective energy, even if everyone doesn't believe in global warming?

NUCLEAR ENERGY

Eleven percent of the electricity generated in the world today is from nuclear energy. Nuclear energy has less than fifteen percent of the carbon footprint of natural gas.[81] The United States is the largest producer of nuclear energy. The United States has one hundred nuclear power plants, followed by France with fifty-eight, and Japan with fifty-one. The PRC has fourteen nuclear power plants with twenty-six more plants under construction. The PRC plans to replace many of their coal-fired powered plants with nuclear reactors.

Nuclear power creates only a small fraction of the carbon footprint of fossil fuel; however, the future for nuclear energy in the world is uncertain. In Europe, countries such as Germany, Belgium, and Lithuania have adopted policies to phase out nuclear power plants. In Asia, countries such as PRC, South Korea, and India have committed to rapid expansion of nuclear power plants.

The future of nuclear power in the United States, United Kingdom, and Japan is murky. The Fukushima Daiichi nuclear disaster in 2011 has caused Japan to rethink

---

80 "A Thousand Barrels a Second" by Peter Tertzakian
81 "Lifecycle Greenhouse Gas Emissions Estimates for Electricity Generators," by Benjamin K. Sovacool

their plans for future nuclear power plants.[82] In the United States and the United Kingdom, numerous studies have shown that the public remains distrustful and uneasy about nuclear power.

In the United States, General Electric has scaled back its nuclear operations over concerns about the economic viability of new nuclear power plants. On March 29, 2017, Westinghouse Electric Company filed for Chapter 11 bankruptcy due to nine billion dollars of losses from its nuclear reactor construction projects.[83]

In the United States, the economics for new nuclear plants are influenced by their capital cost, which accounts for at least 60 percent of the cost of electricity.[84] Cost estimates to build a new nuclear reactor have more than tripled since 2005.[85] New nuclear power plants face high financing costs due to rapid escalation in project costs. New nuclear power plants aren't financially competitive with low-cost fuel sources like natural gas and wind power.

Nuclear power is forecast to decline marginally in Europe over the next ten years. However, nuclear power is forecast to significantly grow in PRC, India, and South Korea over a similar period. Nuclear power in the United States has plateaued due to cost and public perception.

The future of nuclear energy will come down to countries making difficult choices. Does the low carbon footprint of nuclear energy offset the high cost and public perception? The governments of PRC, India, and South Korea have set a clear path toward increasing nuclear power plants in their countries. The governments of the United States and the United Kingdom have yet to establish a firm position on nuclear energy. The global future of nuclear energy over the next thirty years is still uncertain.

RENEWABLE ENERGY

Twenty-two percent of the electricity generated in the world today is from renewable energy. Renewable energy includes wind, hydroelectric, solar, geothermal, and biomass. **Wind energy** has the lowest median greenhouse emissions with 11 to 12 grams CO_2/kWh, followed by **hydroelectric energy** with twenty-four grams CO_2/kWh, **solar energy** with 27 to 48 grams CO_2/kWh, **geothermal energy** with 38 grams CO_2/kWh, and **biomass** with 230 grams CO_2/kWh.[86] As a comparison, the median greenhouse emissions from **coal** is 820 grams CO_2/kWh.

82   "Six Years after Fukushima, much of Japan has lost faith in nuclear power," by Dr. T. Suzuki – March 9, 2017
83   "Westinghouse Bankruptcy Shakes the Nuclear World" by J. Conca, Forbes – March 31, 2017
84   "The Economics of Nuclear Power", World Nuclear Association – June 2017
85   "The High Cost of Nuclear Power" by T. Madsen, J. Neumann and E. Rush – March 31, 2009
86   "Lifecycle Greenhouse Gas Emissions Estimates for Electricity Generators," by Benjamin K. Sovacool

Hydroelectric energy is the dominant renewable energy source in the world, supplying 71% of all renewable electricity. In 2015, hydroelectric power stations generated approximately 3,969,115 gigawatt hours (GWh) of electricity in the world.[87] The PRC generated the most hydroelectric power with 1,126,400 GWh, followed by Brazil with 382,058 GWh, and United States with 250,148 GWh. Significant new hydroelectric projects are planned in the PRC, Latin America, and Africa.

Massive hydroelectric projects in the PRC were the primary cause for the recent growth in global hydroelectric power capacity as shown in *Figure 59*. There is significant undeveloped hydroelectric power potential in Asia, Africa, South America, and North America.

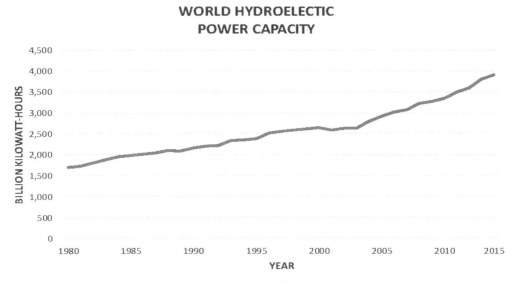

WORLD HYDROELECTIC POWER CAPACITY

Figure 59

The average cost of electricity from a hydroelectric station larger than ten megawatts is $0.046 per kilowatt-hour.[88] Hydroelectricity is a cost-effective, renewable resource with a low carbon footprint. However, large hydroelectric power stations result in the flooding of extensive areas upstream of the dam, which negatively impact local ecosystems. To address climate change, difficult choices will have to be made between local ecosystems and renewable energy, like hydroelectric power.

Solar energy is generated from either a photovoltaic (PV) system or a concentrated solar power (CSP) system. PV systems use solar panels to convert sunlight directly

---

87 "World Energy Resources Hydropower 2016, by World Energy Council – 2016
88 World Energy Council Resource – 2016 Summary

into electricity. CSP systems use solar thermal energy to create steam, which powers turbines that produce electricity. Approximately 98% of the world's solar power is from PV systems.

The largest PV power plant in the world is the 850 MW Longyangxia Dam Solar Park in the PRC. The largest CSP power plant in the world is the 377 MW Ivanpah Solar Power Facility in the United States.

In 2015, solar power generated approximately 253 terawatt-hours (TWh) of electricity in the world, as shown in *Figure 60*.[89] The PRC generated the most electricity from solar energy with 78,100 megawatts (MW), followed by Japan with 42,800 MW, Germany with 41,200 MW, and United States with 40,300 MW.

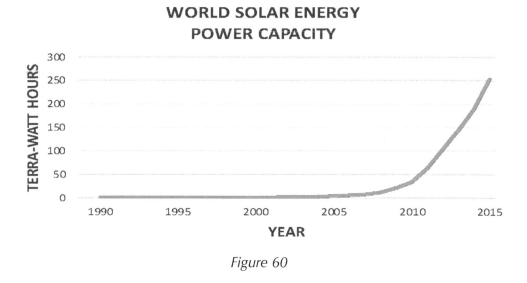

WORLD SOLAR ENERGY POWER CAPACITY

Figure 60

Solar energy is one of the fastest growing forms of renewable energy in the world. The dramatic increase in the use of solar energy over the past ten years has been driven by significant price reductions in the cost of PV systems and tax incentives in some countries. The average cost of electricity from solar PV systems is $0.126 per kilowatt-hour, depending on the available sunlight in the region.[90]

Solar energy has caused the utility industry to rethink the long-held business model of centralized power generation systems. PV systems are causing utilities in the United States to develop new relationships with their own customers and technology developers. In Hawaii, customers can turn extra energy from their own PV system back into the electric grid, earning a credit on their electric bill. This has the

89   BP-Statistical Review of World Energy - 2016
90   World Energy Council Resource, 2016 Summary

potential to revolutionize the power grid system and the role of the utility companies in the United States.

Geothermal energy generates approximately 13.3 gigawatts (GW) of electricity per year in the world.[91] The United States generates the most electricity from geothermal energy with 3,567 megawatts (MW), followed by the Philippines with 1,930 MW and Indonesia with 1,375 MW. The average cost of electricity from geothermal energy is $0.08 per kilowatt-hour.[92]

New geothermal projects are planned in Indonesia, United States, Turkey, Kenya, and Ethiopia. Electricity from geothermal power plants are forecast to grow at a rate of 4% to 5% per year over the next ten years.

Wind energy generated approximately 487 gigawatts (GW) of electricity in the world in 2016.[93] PRC generated the most electricity from wind energy with 169 GW, followed by the United States with 82 GW and the Federal Republic of Germany with 50 GW. Future major wind projects are planned in Canada, Denmark, India, Ireland, Portugal, PRC, United States, and Uruguay. Electricity from wind energy is forecast to grow at a rate of 8% to 10% per year over the next ten years.

Wind power is one of the fastest growing forms of renewable energy in the world, as shown in *Figure 61*. The dramatic increase in the use of wind energy over the past fifteen years has been driven by improvements in turbine technology, reductions in the cost of turbines, and tax incentives in many countries.

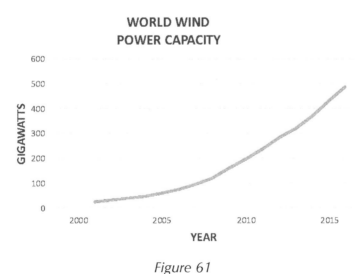

Figure 61

91   "2016 Annual & Global Geothermal Power Production Report by Geothermal Energy Association," March 2016
92   World Energy Council Resource, 2016 Summary
93   "Renewables 2016," Global Statistics Report

Winds offshore tend to blow with more intensity and more uniformly than on land. The potential energy produced from wind is directly proportional to the cube of the wind speed. A small increase in the wind speed can produce a significantly greater amount of electricity. As an example, a wind turbine can produce 50% more electricity with an average wind speed of sixteen miles per hour compared to an average wind speed of fourteen miles per hour.

The higher offshore wind intensity has resulted in the development of offshore wind farms. Forty-seven major offshore wind farms are operating in Europe, with nineteen in Germany, nineteen in the United Kingdom, five in Denmark, three in Belgium, and one in the Netherlands. Three offshore wind farms are operating the in the People's Republic of China (PRC). One offshore wind farm began operation off the coast of Rhode Island in December 2016. The states of Connecticut, Delaware, Maryland, Massachusetts, New York, New Jersey, Ohio, and Virginia are now pursuing offshore wind farm projects.

The average cost of electricity from onshore wind farms is $0.06 per kilowatt-hour. The average cost of electricity from offshore wind farms is $0.159 per kilowatt-hour.[94] Although winds are more intense and consistent offshore, the cost to install a wind turbine offshore is significantly higher than on land. However, companies are now developing floating offshore wind turbines, which has the potential to significantly reduce the electricity costs for offshore wind farms.

Biomass is defined by the European Union (EU) as the biodegradable fraction of products, waste, and residues of biological origin from agriculture, forestry, and related industries, including fisheries and aquaculture, as well as the biodegradable fraction of industrial and municipal waste. Biomass is used in electrical power generation, heating, and transportation. Although biomass produces significant amounts of greenhouse gases, the European Union (EU) and the United Nations (UN) have classified biomass as a renewable energy.

Biomass generated approximately 464,000 gigawatts (GW) of electricity in the world in 2016.[95] The United States generated the most electricity from biomass energy with 69,000 GW, followed by Germany with 50,000 GW and the PRC with 48,000 GW. The United States is also the largest producer of biomass fuel. The average cost of electricity from biomass is $0.055 per kilowatt-hour.[96]

Electricity from biomass energy is forecast to grow at a rate of 2% to 5% per year over the next ten years. Many oppose using biomass, because using trees and

---

94   World Energy Council Resource, 2016 Summary
95   "Renewables 2016," Global Statistics Report
96   World Energy Council Resource, 2016 Summary

crops for fuel will ultimately increase food prices and increase deforestation in the world.

The world is facing an energy conundrum. Will the world continue the unabated consumption of the dwindling supply of fossil fuels? Is the world prepared to accept 3.7 million deaths a year from chronic or acute effects of atmospheric pollutants[97] or gamble that global warming is a myth? Logic would dictate that the world leaders would aggressively push to reduce the use of fossil fuel and increase the use of sustainable, reliable, and cost-effective renewable energy. Time will tell how mankind will deal with the energy conundrum with which we are all face.

97　"Enhancing the World Bank's Approach to Air Quality Management," World Bank, February 2015.

GLOSSARY OF TERMS

American Association of Petroleum Geologists (AAPG): A nonprofit professional organization whose mission is to promote the science of geology, as it pertains to exploring, evaluating, and producing petroleum, natural gas, and other subsurface fluids.

Apache Corporation: In 1954, the Apache Oil Corporation was founded in Minneapolis, Minnesota, by Truman Anderson, Raymond Plank and Charles Arnao. Apache Corporation is an independent oil and natural gas and production company headquartered in Houston, Texas. In 2016, the company's total production was 521 thousand barrels of oil equivalent per day, of which 43% was in the United States, 11% was in Canada, 33% was in Egypt, and 13% was in the North Sea.

Asset Management: In the oil and gas industry, an asset usually refers to producing oil and/or gas fields. Asset management focuses on maximizing the financial performance of the producing fields. Maximizing financial performance can be achieved by improving the rate of oil and gas.

Barrel of Oil: In the oil industry, one barrel of oil is defined as forty-two U.S. gallons or thirty-five imperial gallons. Oil companies usually report their production in terms of volume and use units such as barrels (bbl), one thousand barrels (Mbbl) or one million barrels (MMbbl).

Barrel of Oil Equivalent (BOE): A unit of energy based on the energy released by burning one barrel (forty-two U.S. gallons or 158.9873 liters) of crude oil. The BOE is used by oil and gas companies in their financial statements as a way of combining oil and natural gas reserves and production into a single measure, although this energy equivalence does not take into account the lower financial value of energy in the form of gas. The Society of Petroleum Engineers (SPE) states 5,658.53 cubic feet of natural gas is equivalent to one BOE. The United States Internal Revenue Service (IRS) defines a BOE as equal to 5.8×10^6 BTU.

Beignet: A pastry that is made from square pieces of dough, fried and covered in powdered sugar. New Orleans is known for this delicious French-Creole pastry.

BGP Inc., China National Petroleum Corporation: An international geophysical service company headquartered in Zhuozhou, People's Republic of China (PRC). BGP's services include seismic acquisition of onshore, marine, and transition zone data, seismic data processing, and interpretation.

Black Gold: An informal name for oil. Oil is black when it flows to the surface and is a precious commodity, like gold.

Blow-Out: The uncontrolled release of oil and/or natural gas due to the failure of the pressure control system and blowout preventer on the rig.

Bottom Hole Assembly (BHA): The lower portion of the drill string. The BHA usually consists of the drill bit, a bit sub, a mud motor, stabilizers, drill collar, heavyweight drill pipe, jarring devises (jars), and crossovers for various thread forms. The BHA must provide force for the bit to break the rock, endure high pressures and/or temperatures, and provide the driller with the ability to directionally control or steer the BHA.[98]

Brunei National Petroleum Company Sendirian Berhad (PetroleumBRUNEI): The company was incorporated in 2002 and is based in Bandar Seri Begawan, Brunei. The company's role is to realize and enhance the value of hydrocarbons to the country by the exploration and development of designated areas and recovery of hydrocarbon resources.

British Thermal Unit (Btu): Is a standard unit of energy that is used in the United States. It represents the amount of thermal energy necessary to raise the temperature of one pound of pure liquid water by 1 degree Fahrenheit at the temperature

---

98   Schlumberger Oil Field Glossary

at which water has its greatest density (39 degrees Fahrenheit).[99] Btu is often used as a quantitative specification for the energy-producing or energy-transferring capability of heating and cooling systems such as furnaces, ovens, refrigerators, and air conditioners.

Cable Tool Drilling Rig: Also known as a percussion drilling rig, is a machine that uses repeated penetrations to pound holes into rock, soil, or cement. The first cable tool drilling rig is reported to have been developed over four thousand years ago in China. The first cable tool drilling rigs were used to drill for water. The first successful oil well in the world was drilled in 1859 with a cable tool drilling rig in Titusville, Pennsylvania.

CO_2/kWh_e: A measurement of life-cycle greenhouse gas emissions which calculates the global-warming potential of electrical energy sources through life-cycle assessment of each energy source. The results are presented in units of global warming potential per unit of electrical energy generated by that source. The scale uses the global warming potential unit, the carbon dioxide equivalent (CO_2e), and the unit of electrical energy, the kilowatt hour (kWh). The goal of such assessments is to cover the full life of the source, from material and fuel mining through construction to operation and waste management.

Concentrated Solar Power (CSP): Uses mirrors to concentrate the energy from the sun to drive traditional steam turbines or engines that create electricity. The thermal energy concentrated in a CSP plant can be stored and used to produce electricity when it is needed, day or night. Today, over 1,800 megawatts (MW) of CSP plants operate in the United States.[100]

Consolidated Contractors Energy Development (CCED): A privately owned Lebanese independent upstream oil and gas exploration and production company. It was established in 2007 with the aim of acquiring exploration and development assets in the Middle East, Africa, and Commonwealth of Independent States (CIS).

Chevron: In 1911, the Supreme Court ordered the dissolution of Standard Oil Company, ruling it was in violation of the Sherman Antitrust Act. Standard Oil Co. (California), was founded because of the dissolution of Standard Oil Company. The

99 WhatIs.com
100 Solar Energy Industries Association

Chevron name came into use for some of its retail products in the 1930s. Chevron is an American multinational energy corporation, headquartered in San Ramon, California, and active in more than 180 countries.

Deepwater: No precise definition has been established for the terms deepwater and ultra-deepwater. Many companies consider water depths from four thousand feet to seven thousand feet to be "deepwater" and water depths greater than seven thousand feet to be "ultra-deepwater."

Debottlenecking: Increasing production capacity of existing facilities through the modification of existing equipment to remove throughput restrictions. The debottlenecking process is used to optimize the performance of power plants and energy distribution systems.[101]

Development Prospect: An undrilled area of a proven field that development geologists, geophysicists, and reservoir engineers believe contains commercial qualities of hydrocarbons. Globally, the probability of a commercially successful development well ranges from 85% to 95%.

Direct Hydrocarbon Indicator (DHI): Anomalous seismic responses caused by the presence of hydrocarbons. DHIs occur when a change in pore fluids causes a change in the elastic properties of the bulk rock that is seismically detectable (i.e., there is a "fluid effect").[102]

Dog House: Oil and gas industry slang for the seismic recording truck, which is used in onshore seismic operations. The seismic recording truck contains various electronic recording, filtering, and timing instruments. An energy source (trucks with vibrators, explosives, etc.) generates seismic waves, which are sent down through the substrate to be reflected in the rock medium back to the surface. The returning signals are picked up by a nearby network of ground microphones (geophones) and then fed into the seismic recording truck.

Drilling Engineer: An engineer that designs and implements procedures to drill wells safely and economically. The drilling engineer works closely with the drilling contractor, service contractors, compliance personnel, geologists, and other technical specialists.

---

101 Encyclo.co.uk
102 Incorporated Research Institutions for Seismology (IRIS), November 1, 2016

Drill Ship: A vessel designed for offshore drilling of oil and gas wells. Over the past ten years, drill ships have been used to drill deepwater and ultra-deepwater wells.

Dry Hole: A drilled well that does not contain commercial qualities of hydrocarbons. A dry hole is sometimes referred to as "a duster."

Exploration Play: Describes a group of undrilled prospects that have similar geological and/or geophysical attributes. As an example, a group of structural traps, which have the same reservoir type and share a common source rock, may be called an exploration play.

Exploration Prospect: An undrilled trap that exploration geologists and geophysicists believe may contain commercial quantities of hydrocarbons. Four geological factors (trap integrity, reservoir quality, presence of source hydrocarbon generating source rocks, and hydrocarbon migration into the trap) must be present for a prospect to contain hydrocarbons. Globally, the probability of an exploration well being commercially successful usually ranges from 20% to 35%.

Farm-In: A contractual agreement whereby one company acquires an interest in a mineral interest by another company. The farm-in can be for opportunities ranging from an undrilled exploration prospect to a producing oil or gas field.

Farm-Out: A contractual agreement whereby one company sells an interest in a mineral interest to another company. A farm-out is sometimes negotiated to mitigate the cost of a higher-risk exploration well or to fund the development of a recently discovered oil and gas field.

Floating Production Storage and Offloading (FPSO): A floating facility that is equipped with hydrocarbon processing equipment for separation and treatment of crude oil, water, and gases, arriving onboard from subsea oil wells via flexible pipelines.

Forest Oil: Forest Oil Corporation is credited with developing the secondary oil recovery technique ("waterflooding") in the early 1900s. Texas. On December 16, 2014, Forest Oil Corporation was acquired by Sabine Oil & Gas Corporation.

Geophones: A device that is placed in the ground that converts ground movement (velocity) into voltage, which is then recorded on the seismic field computer. The deviation of this measured voltage from the base line is called the seismic response. The seismic responses are then processed, and the output data is analyzed to interpret the subsurface structure and stratigraphy of the earth.

Geophysicist: A scientist who uses physics, chemistry, geology, and advanced mathematics to study the subsurface characteristics and composition of the Earth. In the oil and gas industry, geophysicists use sophisticated instruments to measure the acoustic, magnetic, and gravimetric properties of the subsurface strata to fine commercial quantities of hydrocarbons.

Geoscience Australia: Agency of the Australian government for geoscience research and geospatial information. Geoscience Australia's research and information provided input for the governments decisions on research use, management of the environment, and the safety and well-being of Australians.[103]

Giant Field: An oil field that contains over five hundred million barrels of recoverable oil and a gas field that contains over three trillion cubic feet of recoverable natural gas.[104]

Gulf of Mexico Bathometric Classification: There is no established criteria to precisely differentiate between shallow water, deepwater, or ultra-deepwater. Many Gulf of Mexico companies consider "shallow water" to be water depths less than 1,000 feet (305 meters), "deepwater" to be water depths between 1,000 feet (305 meters) and 5,000 feet (1,524 meters), and ultra-deepwater to be water depths greater than 5,000 feet (1,524 meters).

Hydraulic Well Fracturing ("Fracking"): This is the process of pumping fluid into a wellbore to create sufficient pressure to fracture the rock layer. The fluid usually contains a "proppant," like sand, that keeps the rock fractures open to allow oil and/or gas to flow to be produced.[105]

Huffco: Roy Michael Huffington (October 4, 1917 – July 11, 2008) established an oil and gas company, known as Roy M. Huffington, Inc. (Huffco) in 1956. In 1972, Huffco discovered the giant natural gas field, Badak in East Kalimantan, Indonesia.

---

103 www.ga.gov.au/about/history
104 AAPG Memoir 40 by Michel T. Halbouty in 1986
105 Alberta Energy Regulator

Igneous Rock: The word igneous is from the Greek word for fire. Igneous rocks are formed when hot, molten rock crystallizes and solidifies. The molten rock originates deep within the Earth near active plate boundaries, or hot spots, then rises toward the surface. Granite, rhyolite, and basalt are types of igneous rocks.[106]

Independent Company: A company that is usually in the exploration and production segment of the industry, with no marketing or refining within their operations. Most independent companies receive their revenue from hydrocarbon production.

International License Round: The process for a sovereign country to award a hydrocarbon license to an oil and gas company (major company, independent company, or national oil company). The procedure varies by country. Most international license rounds require companies to submit sealed bids of work programs (the number of exploration wells, miles of 2-D seismic data, studies, etc.) to be conducted over a finite period of time. The company with the most significant work program is usually awarded the hydrocarbon license.

Interpretation Workstation: A desktop computer system that allows the interpreter to integrate and interpret all available digital geological, geophysical, etc. data. The interpreter will generate various three-dimensional maps of prospective horizons. Interpretation workstations are used to assess a diversity of opportunities ranging from exploration prospects to the characterization of producing reservoirs in oil or gas fields.

Jack-Up Drilling Rig: The drilling rig consists of a buoyant hull, which is fitted with moveable legs, capable of raising the hull above the surface of the sea. Tugs or heavy lift ships tow the rig to the drilling location. Once the rig is on location, the moveable legs raise the hull to the designated elevation above the sea surface. After the well has been drilled, the moveable legs lower the hull into the water, and the rig is towed to the next drilling location.

Jug Hustler: Oil and gas industry slang for a member of a seismic acquisition crew who lays out cables and plants geophones in the ground prior to the seismic data acquisition. The jug hustler then picks up the geophones and cables once the seismic line has been acquired.

---

106 United States Geological Survey

Korea National Oil Company (KNOC): The national oil company of the Republic of Korea (South Korea). KNOC was founded in 1979 to stabilize the supply of energy through strategic petroleum stockpiling and development. The company operates in six segments: oil and gas, petroleum distribution, oil stockpiling, financing, drillship chartering, and others.

Landman: In the United States or Canada, the roles for a petroleum landman include determining the ownership of mineral rights through the research of public and private records, reviewing the status of title, curing title defects, and otherwise reducing title risk associated with ownership in minerals and managing rights and/or obligations derived from ownership of interests in minerals and negotiation for the acquisition or divestiture of mineral right.

Liquefied Natural Gas (LNG): Natural gas which has been cooled to the point of liquefaction. LNG is odorless, colorless, noncorrosive, and nontoxic. LNG is created when natural gas is cooled to -260 degrees Fahrenheit.

Logging While Drilling (LWD): Tools integrated into the bottomhole assembly (BHA) that can measure the properties of the subsurface formation during the drilling of a well.[107]

Major Company: A company with significant upstream and downstream operations. The term is occasionally used to refer to the six largest major operators (BP, Chevron, Eni SpA, ExxonMobil, Shell, and Total). The term "major company" doesn't include national or state-owned oil companies, such as ARAMCO – Saudi Arabia, NIOC – Iran, PEMEX – Mexico, CNPC – China, etc.

Marine-Riser System: A large diameter pipe that connects the subsea blowout preventer (BOP) stack to a floating surface rig to take mud returns to the surface. Without the riser, the mud would simply spill out of the top of the stack onto the seafloor.[108]

Market Capitalization: It is the stock market value of a company's outstanding shares.[109] A company's market capitalization is calculated by taking the stock price and multiplying it by the total number of outstanding shares.

---

107 Schlumberger Oil Field Glossary
108 Schlumberger Oil Field Glossary
109 Investopedia

Metamorphic Rock: A metamorphic rock is formed when either an igneous or sedimentary rock is subjected to high heat, high pressure, hot, mineral-rich fluids or some combination of these factors. Marble, quartzite, slate, and schist are types of metamorphic rocks.[110]

Mineral Management Service: The Minerals Management Service (MMS) was an agency of the United States Department of the Interior that managed the nation's natural gas, oil, and other mineral resources on the outer continental shelf (OCS).

As a result of the Deepwater Horizon oil spill, Secretary of the Interior Ken Salazar issued a secretarial order on May 19, 2010 splitting MMS into three new federal agencies: the Bureau of Ocean Energy Management, the Bureau of Safety and Environmental Enforcement, and the Office of Natural Resources Revenue. MMS was temporarily renamed the Bureau of Ocean Energy Management, Regulation and Enforcement (BOEMRE) during this reorganization before being formally dissolved on October 1, 2011.

Ministry of Oil and Gas (MOG): The government body in the Sultanate of Oman responsible for developing and implementing the government policies for exploiting the oil and gas resources in the country.

Mitsui Exploration and Production Middle East B.V.: The company is based in Amsterdam, the Netherlands, and explores, develops, and produces oil and natural gas in the Sultanate of Oman.[111]

Mobil Oil: In 1911, the Supreme Court ordered the dissolution of Standard Oil Company, ruling it was in violation of the Sherman Antitrust Act. Standard Oil Company of New Jersey, or Socony, was founded because of the dissolution of Standard Oil Company. In 1920, the company registered the name "Mobiloil" as a trademark. In 1999, Exxon and Mobil merged, forming the company called ExxonMobil.

Mud Logging: Also known as hydrocarbon well logging, mud logging is the gathering of qualitative and quantitative data from hydrocarbon gas detectors that record the levels of natural gas brought up in the drilling mud. Chromatographs are used to determine the chemical composition of the gas. Mud logging also includes recording

---

110 United States Geological Survey
111 Bloomberg

the drilling rate, mud weight of the drilling mud, rock type, or lithology of the drill cuttings.[112]

Murphy Oil: Murphy Oil Corporation developed from family timberlands in southern Arkansas and northern Louisiana that were owned by Charles H. Murphy Sr. The company was officially formed in 1950 by the children of Charles H. Murphy Sr. The Murphy Oil Corporation now operates oil and gas production facilities and processing plants across the world.[113]

Net Present Value (NPV): Is the difference between the present value of cash inflows and the present value of cash outflows. NPV is routinely calculated for all projects to determine capital allocation.

Oceania: The region that lies between the continents of Asia and North America. Oceania consists of countries and territories in the Pacific Ocean, including Australia, East Timor, and New Zealand.

Outer Continental Shelf (OCS): Is governed by Title 43, Chapter 29, "Submerged Lands, Subchapter III, Outer Continental Shelf Lands," of the U.S. Code.[114] The term "outer Continental Shelf" refers to all submerged land, its subsoil, and seabed that belong to the United States and are lying seaward and outside the states' jurisdiction, the latter defined as the "lands beneath navigable waters" in Title 43, Chapter 29, Subchapter I, Section 1301.

Petroleum Development Oman (PDO): The company is owned by the Government of Oman (60%), Royal Dutch Shell (34%), Total (4%) and Partex (2%). PDO's primary objective is to engage efficiently, responsibly, and safely in the exploration, production, development, storage, and transportation of hydrocarbons in the Sultanate of Oman. PDO operates in a concession area of about ninety thousand square kilometers (one third of Oman's geographical area).[115]

Petroleum Geologist: An earth scientist who is involved in the search for commercial oil and gas fields. Petroleum geologists look at the stratigraphic and structural aspects of the subsurface strata to identify possible oil traps.

---

112 Crain's Petrophysical Handbook
113 The Encyclopedia of Arkansas History and Culture
114 Bureau of Ocean Energy Management
115 Petroleum Development Oman, pdo.co.om

Petrophysicist: A scientist who is involved in the study of physical and chemical rock properties and their interactions with fluids, utilizing electric logs, physical rock, and fluid measurements.

PETRONAS Twin Towers: Two skyscraper office buildings in Kuala Lumpur, Malaysia, that are among the tallest office buildings in the world. The office buildings were designed by the American architect Cesar Pelli and completed in 1998.

Photovoltaic (PV) Cell: Commonly called a solar cell, this is a nonmechanical device that converts sunlight directly into electricity. Some types of photovoltaic cells can convert artificial light into electricity.[116]

Prime Interest Rate: The prime rate is the interest rate that commercial banks charge their most credit-worthy customers, who are generally large corporations. The prime interest rate, or prime-lending rate, is primarily determined by the federal funds rate, which is the overnight rate that banks use to lend to one another. The prime rate also impacts individual borrowers, as the prime rate directly affects the lending rates for a mortgage, small business loan, or personal loan.

Production Engineer: An engineer that designs and selects subsurface equipment to produce oil and gas well fluids. Production engineering is considered a subset of petroleum engineering.

Promote: The premium the investor pays to participate in an opportunity, usually a drillable prospect. The "promote" usually applies to the entire cost of drilling and to completing the first well on a prospect. The promote premium depends on many factors, including prospect quality, hydrocarbon price, competition, etc.

PTT Exploration and Production Public Company Limited (PTEP): Thailand's petroleum exploration and production company, which was established in June 1985. PTTEP is a subsidiary of the Petroleum Authority of Thailand.

PETRONAS Carigali Sdn. Bdh.: Is a wholly owned exploration and production subsidiary of PETRONAS, the national oil company of Malaysia. The company is actively exploring and producing hydrocarbons in over thirty countries in the world.

---

116 U.S. Energy Information Agency

Proven Reserves: Denotes the amount of oil and/or gas that can be recovered at a cost that is considered economic at the present market price of the hydrocarbon. Proven reserves will change with the hydrocarbon price.

Return on Capital Employed (ROCE): The ROCE represents total assets, less current liabilities, and is used to show how much a business is gaining for its assets.

Rate of Return (ROR): The profit or loss on an investment over a period of time, expressed as a percentage of the original investment. In the oil and gas industry, the period is usually determined by length of time to start and complete an operation, such as drilling a well.

Security Exchange Commission: The United States Securities and Exchange Commission (SEC) is an independent federal government agency responsible for protecting investors, maintaining fair and orderly functioning of securities markets, and facilitating capital formation. It was created by Congress in 1934 as the first federal regulator of securities markets. The SEC promotes full public disclosure, protects investors against fraudulent and manipulative practices in the market, and monitors corporate takeover actions in the United States.

Sedimentary Rock: Existing rock that is weathered then transported to a depression or basin, where the rock fragments or sediment is deposited. If the sediment is deeply buried, it becomes compacted and cemented, forming sedimentary rock. Limestone, sandstone, siltstone, and shale are types of sedimentary rocks.[117]

Service Company: A company that provides products and/or services to the oil and gas industry. Services include drilling, logging, well testing, well completion, seismic acquisition, etc. Compagnie Générale de Géophysique (CGG), Diamond Offshore Drilling, Halliburton, Nabors Industries, Transocean Drilling, and Schlumberger are examples of service companies in the oil and gas industry.

Shale Oil: A type of unconventional oil found in shale formations. Shale oil can refer to oil that is found within shale formations or to oil that is extracted from oil shale. Oil shale has bituminous-like solids that can be liquefied during the extraction process.[118]

---

117 United States Geological Survey
118 Investopedia

Shale Gas: Natural gas that is trapped within shale formations. Horizontal drilling and hydraulic fracturing has allowed access to large volumes of shale gas that were previously uneconomical to produce. The production of natural gas from shale formations has rejuvenated the natural gas industry in the United States.[119]

State Waters: The federal government administers the submerged lands, subsoil, and seabed, lying between the seaward extent of the states' jurisdiction and the seaward extent of federal jurisdiction.[120] The jurisdiction for the State of Texas is extended nine nautical miles seaward from the baseline from which the breadth of the territorial sea is measured. The jurisdiction for the State of Louisiana is extended three U.S. nautical miles (U.S. nautical mile = 6080.2 feet) seaward of the baseline from which the breadth of the territorial sea is measured. All other states' seaward limits are extended three International Nautical Miles (International Nautical Miles = 6076.10333 Ft.) seaward of the baseline from which the breadth of the territorial sea is measured.

Reservoir Engineer: An engineer that applies scientific principles to optimize the hydrocarbon recovery of subsurface reservoirs during the development and production of fields. One of the primary responsibilities of the reservoir engineer is the accurate assessment of reserves estimates for financial reporting to regulatory organizations, such as the United States Securities and Exchange Commission (SEC).

Scout: A professional in the oil and gas industry that can unearth valuable information about competitors' activities. A scout must have a basic knowledge about geology, land leasing, well testing, well logging, and current oil and gas regulations.

Seismic Acquisition: The generation and recording of seismic data onshore or offshore. The acquisition of seismic data involves many different types of receiver configuration, such as laying geophones in the surface of the ground or on the seafloor or towing hydrophones behind a marine vessel. An energy source, such as a vibrator unit or an air gun, generates the acoustic vibrations that travel into the subsurface and are reflected back to the surface.

Seismic Processing: Seismic data that is acquired onshore or offshore must be processed to be in a format that a geophysicist can interpret on the workstation. Coherent

---

119 Geology.com
120 Bureau of Ocean Energy Management

and ambient noise will be removed, and the real seismic data enhanced in the seismic processing center.

2-D Seismic Data: Provides a two-dimensional perspective of the subsurface under the seismic line. However, two-dimensional seismic data contains a mixture of both in-plane and out-of-plane reflections. Two-dimensional seismic data usually costs significantly less than three-dimensional seismic data.

3-D Seismic Data: Provides a three-dimensional perspective of the subsurface under the survey area. The location of the reflections is more accurately determined than two-dimensional seismic data. Value of information analysis is routinely used to determine whether the higher cost, but significantly higher quality three-dimensional seismic data should be acquired around interest.

Semisubmersible Drilling Rig: The submersible drilling platform is supported on large pontoon-like structures. These pontoons provide buoyancy, allowing the unit to be towed from location to location. Once on the location, the pontoon structure is slowly flooded until it rests securely on its anchors. After the well is drilled, the water is pumped out of the buoyancy tanks and the vessel is refloated and towed to the next location.

Shallow Hazard Assessment: A predrill study to identify potential drilling hazards, such as abnormal pressure zones, shallow gas, seafloor stability, shallow water flow, and gas hydrates. The study may result in the surface location of the well being moved to mitigate the drilling risk of a potential shallow hazard.

Short Ton: A short ton is a unit of weight equal to 2,000 pounds or 907.18474 kilograms. In the United States, a short ton is known simply as a ton.

Side Tracking a Well: The drilling of a secondary wellbore, away from the original hole. It is possible to drill multiple sidetracks, each of which might be drilled for a different reason. A sidetrack operation may be conducted to reach a different target or to bypass irretrievable junk in the hole or a collapsed wellbore.[121]

Society of Exploration Geophysicists (SEG): This is a nonprofit professional organization whose mission is to promote the science of geophysics and the education of

---

121 Schlumberger Oil Field Glossary

exploration geophysicists. The organization promotes the expert and ethical practice of geophysics in the exploration and development of natural resources and the mitigation of earth hazards.

Society of Petroleum Engineers (SPE): This is a nonprofit professional organization whose mission is to collect, disseminate, and exchange technical knowledge concerning the exploration, development, and production of oil and gas reservoirs for public benefit. The organization also provides opportunities to develop technical and professional skills and expertise.

Spar: A type of floating oil platform typically used in very deep waters. The term is named for logs that are moored in place vertically, used as buoys in shipping. The deep draft design of spars makes them less affected by wind, wave, and currents, and allows for both dry tree and subsea production.

Speculative Seismic Survey: A seismic survey which is acquired and processed by a seismic contractor (WesternGeco, CGG, etc.) and then licensed for a fee on the open market. The seismic contractor owns the exclusive rights to the seismic data. A company pays a licensing fee to use the seismic data for a finite period. Speculative seismic surveys are also known as nonexclusive or multiclient geophysical data licenses.

Standard Cubic Foot (SCF): One cubic foot of gas at standard temperature (60 degrees Fahrenheit) and pressure (Sea Level). SCF is a unit designed to standardize the measurement of natural gas. One SCF is equivalent to 1020 British Thermal Units (BTUs). One MCF is equivalent to one thousand cubic feet of gas.

Terawatt-Hour (TWh): A unit of energy equal to 3.6×10^{15} joules or 1,000 gigawatt hours (GWh). The terawatt hour is used for metering large quantities of electrical energy.

Tethys Oil: A Swedish energy company focused on exploration and production of oil. The company's core area is the Sultanate of Oman.[122]

Ultra-Deepwater: No precise definition has been established for the terms "deepwater" and "ultra-deepwater." Many companies consider water depths from four thousand feet to seven thousand feet to be "deepwater" and water depths greater than seven thousand feet to be "ultra-deepwater."

---

122 www.tethysoil.com/en

United States Federal Offshore Leasing: The state governments own the offshore mineral rights out to three nautical miles from the shoreline. The federal government owns the offshore mineral rights beyond the states' three-mile nautical mile limit. The federal government holds mineral rights lease sales for the Outer Continental Shelf (OCS) annually. Companies submit sealed cash bids for the block(s) of interest. The highest bidder is awarded the mineral rights for a finite period. If the companies do not establish hydrocarbon production within the contract period, then the mineral rights revert to the government and the lease will be retendered at future lease sales.

Vibroseis: A device that uses a truck-mounted vibrator plate coupled to the ground to generate a wave train up to seven seconds in duration and comprising a sweep of frequencies. The recorded data from an increasing frequency or decreasing frequency are added together and compared with the source input signals to produce a conventional-looking seismic section.

Water Injection Well: A well that injects water into the reservoir to increase the formation pressure and enhance the oil recovery. In the Farha South Field in the Sultanate of Oman only 15% of the oil in place in the reservoir can recovered without water injection wells. However, the use of water injection wells, also known as water flooding, has increased the recovery of the oil in place from 15% to over 35%.

Wave Energy: Relies on the up-and-down motion of waves to generate electricity. Experimental installations have been built in Australia, India, Japan, PRC, Norway, and the United States.

Well Test: The execution of a data acquisition plan to broaden the knowledge and understanding of the hydrocarbon properties and characteristics of the subsurface reservoir. The well test objective is to identify the reservoir's capacity to produce economic quantities of hydrocarbons.

Wellsite Geologist: A geologist who performs specialized tests and analysis at the drilling location or wellsite. The wellsite geologist works closely with the mudloggers and drilling engineers to ensure the well is drilled efficiently and safely. The wellsite geologist uses diverse data types, including rock cuttings, well logs, and core sample to assess the geological formation penetrated by the drill bit compared to the predrill prediction.

Wet Gas: Natural gas that contains methane (typically less than 85 percent methane) and more ethane and other more complex hydrocarbons. Ethane and butane are natural gas liquids and can be separated and sold on their own.[123]

Wireline Logging: The process of measuring various rock and fluid properties in a drilled well. Wireline logging is used to make critical decisions about future drilling and/or hydrocarbon operations in the well. In the wireline logging operation, a sonde is gradually put down the borehole and then slowly raised to measure the properties of the rock formations and the fluid properties in the rock formations.

---

123 Schlumberger Oil Field Glossary

REFERENCES

1   AAPG Memoir 40, Michel T. Halbouty, 1986

2   "A Thousand Barrels a Second," Peter Tertzakian

3   American Wind Energy Association

4   Annual U.S. & Global Geothermal Power Production Report, Geothermal Energy Association, March 2016

5   BP-Statistical Review of World Energy, 2016

6   Bureau of Ocean Energy Management

7   Crain's Petrophysical Handbook

8   Dictionary of Modern Politics of Southeast Asia, J. Liow and M. Leifer, 2014

9   Durham University Centre for Borders Research, 2009

10   Environmental & Engineering Geophysical Society

11   "Evolution & Revolution of the E&P Industry," Jack Kerfoot, 2006 International AAPG, Perth Australia

12   Fortune Global 500, 2016

13   Geoscience Australia

14   Geothermal Energy Association

15   Gulf of Mexico Foundation

16   "Historical Oil Price Chart," Tim McMahon, October 6, 2016

17   "Historical Statistics of the United States," U.S. Department of Commerce, 1957

18   Incorporated Research Institutions for Seismology (IRIS), November 1, 2016

19   "Lifecycle Greenhouse Gas Emissions Estimates for Electricity Generators," Benjamin K. Sovacool

20   Murphy Oil Corporation 10-K Form, 2002 through 2011

21   "Basic Materials Observer," Morningstar, April 2014

22   "Natural Gas in China a Regional Analysis," The Oxford Institute for Energy Studies, November 2015

23   NOAA National Climate Data Center

24   "Nuclear Follies," James Cook, Forbes Magazine, February 11, 1985

25   "One-Third of our Greenhouse Gas Emissions Come from Agriculture," Nature, October 31, 2012

26   Petroleum Development Oman, pdo.co.om

27   "Petroleum Products Supplied by Type," DOE, EIA Annual Energy Review, 2008

28   Quandl.com/topics/mur-market-cap

29   Quantifying the Uncertainty in Forecasts of Anthropogenic Climate Change, *Nature*, October 5, 2000.

30   Renewable Energy World

31   Society of Petroleum Engineers (SPE)

32   Schlumberger Oil Field Glossary

33   *Shell Shock: The Secrets and Spin of an Oil Giant*, Ian Cummins and John Beasant

34   "Six Years After Fukushima, Much of Japan Has Lost Faith in Nuclear Power," Dr. T. Suzuki, March 9, 2017

35   Solar Market Industry Association

36   *The Atomic Age*, David Dietz, 1945

37   "Trend of Coal Exports," eaber.org/node/2200

38   U.S. Census Bureau

39   U.S. Department of the Interior Minerals Management Service Gulf of Mexico Region

40   U.S. Department of Labor, Mine Safety and Health Administration, Form 7000-2, Quarterly Mine Employment and Coal Production Report.

41   U.S. Energy Information Administration

42   U.S. Geological Survey

43   U.S. Office of Energy Efficiency & Renewable Energy, March 7, 2016

44   "World Population Growth," Max Roser and Esteban Ortiz-Ospina

45   World Weather & Climate Information

46   www.livefrombeijing.com/2009/11

47   www.mog.gov.om/

48   www.nasa.gov

49   www.omansultanate.com/politics.htm

50   www.tethysoil.com/en